CAROLYN STEEL

Carolyn Steel is a leading thinker on food and cities. Her first book, *Hungry City*, received international acclaim, establishing her as an influential voice in a wide variety of fields across academia, industry and the arts. It won the Royal Society of Literature Jerwood Award for Non-Fiction and was chosen as a BBC Food Programme book of the year. A London-based architect, academic and writer, Carolyn has lectured at the University of Cambridge, London Metropolitan University, Wageningen University and the London School of Economics, and is in international demand as a speaker.

carolynsteel.com
@carolynsteel

ALSO BY CAROLYN STEEL

Hungry City

CAROLYN STEEL

Sitopia

How Food Can Save the World

VINTAGE

1 3 5 7 9 10 8 6 4 2

Vintage is part of the Penguin Random House group of companies
whose addresses can be found at global.penguinrandomhouse.com

Penguin
Random House
UK

Copyright © Carolyn Steel 2020

Carolyn Steel has asserted her right to be identified as the
author of this Work in accordance with the Copyright,
Designs and Patents Act 1988

First published in Vintage in 2021
First published in hardback by Chatto & Windus in 2020

penguin.co.uk/vintage

Front jacket: adaptation of the definition of 'Utopia' from
The Concise Oxford Dictionary of Current English (Oxford, 1984)
used by permission of Oxford University Press. Reproduced by
permission of the Licensor through PLSclear.

A CIP catalogue record for this book is available
from the British Library

ISBN 9780099590132

Printed and bound in Great Britain by Clays Ltd, Elcograf S.p.A.

The authorised representative in the EEA is Penguin Random House
Ireland, Morrison Chambers, 32 Nassau Street, Dublin D02 YH68.

Penguin Random House is committed to a sustainable future
for our business, our readers and our planet. This book is
made from Forest Stewardship Council® certified paper.

MIX
Paper from
responsible sources
FSC
www.fsc.org
FSC® C018179

In loving memory of my mother and father,
who always fed me well

Contents

Introduction

Several years ago, I attended a TEDGlobal conference in Edinburgh. On the last day, after nearly a week spent listening to dozens of thinkers, inventors, artists and activists all talking about their inspiring lives and work, I was slumped exhausted on a beanbag when a tall Dutchman approached me and introduced himself as a senior director of Shell. 'I'm looking for answers,' he said. 'I've been here all week listening to people and have heard nothing important. We have vast problems to solve here! Have you got any good ideas? If you can give me one, I have millions to invest!'

After several days absorbing what had felt to me like a non-stop stream of good ideas, I was somewhat taken aback. Nevertheless, I reflected on what the man from Shell had said and eventually told him that what I thought we most lacked in the world was philosophy. 'We've forgotten how to ask the big questions,' I said, 'such as what makes a good life.' I'll never forget the look on his face. It went from incomprehension and incredulity to impatience and, finally, anger. '*We don't have time for that!*' he almost spat at me. 'We are seven billion people, living beyond our means, destroying the planet, and you say that *what we need is philosophy?*'

Although not the immediate inspiration for this book, this exchange did help to galvanise my reasons for writing it. As the highly stressed Dutchman pointed out, we twenty-first-century humans find ourselves facing multiple life-threatening challenges; ones that require big thinking, urgent action and global cooperation if we are to sort them out. On that, the oilman and I were heartily agreed; where we differed was in our approach to tackling the crisis. Whereas he sought technical solutions to our various problems, I wanted to address their underlying causes by examining the factors, assumptions and choices that had created them. While technology and philosophy are hardly mutually exclusive disciplines – clearly we need both – what our grumpy exchange on the beanbags demonstrated was the gulf that can exist between the two. It is this divide that I seek to bridge in this book through the medium of food.

Why food? Because it is by far the most powerful medium available to us for thinking and acting together to change the world for the better. Food has shaped our bodies, habits, societies and environments since long before our ancestors were human. Its effects are so widespread and profound that most of us can't even see them, yet it is as familiar to us as our own face. Food is the great connector, the stuff of life and its readiest metaphor. It is this capacity to span worlds and ideas that gives food its unparalleled power. It is, you might say, the most potent tool for transforming our lives that we never knew we had.

In my first book *Hungry City*, I explored how the feeding of cities has shaped civilisations over time. The book followed food's journey from land and sea via road and rail to market, kitchen, table and waste dump, showing how each stage of the journey had shaped people's lives around the world. By the end of writing the book, I had come to realise quite how profoundly food shapes virtually every aspect of our existence. I decided to call the last chapter 'Sitopia' (from the Greek *sitos*, food + *topos*, place), in order to name the phenomenon that I'd discovered: the fact that we live in a world shaped by food. In some ways, food's influence is obvious (when we're hungry, for example, or when we can't do up our trousers), yet in other ways its effects are deep and mysterious. How many of us stop to wonder, for example, about food's influence over our minds, values, laws, economies, homes, cities and landscapes – even our attitudes towards life and death?

This book follows on from that earlier discovery. Food shapes our lives, yet since its influence is too big to see, most of us are unaware of the fact. We no longer value food in the industrialised world, paying as little for it as possible. As a result, we live in a bad sitopia, in which food's effects are largely malign. Many of our greatest challenges – climate change, mass extinction, deforestation, soil erosion, water depletion, declining fish stocks, pollution, antibiotic resistance and diet-related disease – stem from our failure to value food. Yet, as this book will argue, by valuing food once again, we can use it as a positive force, not only to address such threats and reverse numerous ills, but to build fairer, more resilient societies and lead happier, healthier lives.

Like *Hungry City*, *Sitopia* is arranged in seven chapters representing a food-based journey, in this case starting with a plate of food and travelling out to the universe. The story begins with food itself, moving out to the body, the home, society, city and country, nature and time. At each stage – or scale – of this journey I use food as a lens to explore the origins and dilemmas of our current situation and to ask how we can improve it.

Food lies at the heart of sitopia, yet this book is not primarily about food; rather it explores how food can help us to address our many quandaries in a connected and positive way. We can't live in utopia, but by thinking and acting through food – by joining forces to build a better sitopia – we can come surprisingly close.

1

Food

Google Burger

Technology is the answer. But what was the question?

Cedric Price[1]

In August 2013, an audience gathered in London to witness a remarkable gastronomic event. Broadcast live from a TV studio and hosted by ITN news anchor Nina Hossain, it involved the cooking and tasting of the world's first lab-grown beefburger. Crackling with tension, the occasion had the incongruous air of a Saturday-morning cookery show hijacked by some secretive research facility. Instead of the usual celebrity guests and breezy chat, there was the burger's creator, Maastricht University Professor of Physiology Mark Post, perched uneasily on a stool, next to two anxious-looking 'guinea pigs' – Austrian nutritionist Hanni Rützler and US food writer Josh Schonwald – ready to try what might be the food of the future.

Revealed from beneath a silver cloche, the burger looked innocent enough, although on closer inspection its purplish hue and too-smooth texture (plus the fact that it sat in a Petri dish) betrayed its unique provenance. Created over the course of five years at a cost of €250,000, the burger consisted of 20,000 strands of what Post called 'cultured beef' – in-vitro muscle tissue grown from bovine stem cells – mixed with some more familiar ingredients: egg and breadcrumbs for texture, plus saffron and beetroot juice for colour. Richard McGeown, the chef charged with cooking this precious puck of protein, scooped it up with the air of a man handling nuclear waste and lowered it gingerly into a pan of melted butter.

As the patty started to sizzle, a short film was shown explaining the science behind in-vitro meat. With cartoony graphics and a jazz-funk soundtrack straight out of the 'Dino-DNA' sequence in *Jurassic Park*, a velvety American baritone informed us that the muscle tissue for cultured beef is initially 'harvested' from a cow in a 'small and harmless procedure'. The fat and muscle cells are then separated and the latter dissected, causing them to self-divide. 'From one muscle cell, more than *one trillion* cells can be grown!' purred the voice. The cells then merge to produce 0.3-millimetre-long chains that are placed around a central hub of gel, where their natural tendency to contract causes them to bulk up, producing more muscle. 'From one small piece of tissue, *one trillion* strands can be produced!' the voice enthused, seemingly unaware of the repetition. 'When all these little pieces of muscle are layered together, we get exactly the same thing we started with: *beef*!'

Back in the studio, Chef McGeown pronounced the burger ready, serving it up on a white plate next to a desultory bun, slice of tomato and a lettuce leaf. 'Ladies first!' chirruped Hossain, pushing the plate towards Rützler, who tentatively cut off a small piece of patty, peered and sniffed at it and then put it into her mouth and began to chew. As this 'one small bite' moment of food history played out, Post explained how Winston Churchill had predicted all this back in 1931, in an essay in which he described how humans would one day 'escape the absurdity' of rearing whole chickens by growing the edible parts in a 'suitable medium'.[2] As Post warmed to his theme, it became clear to everyone else that the burger was burning Rützler's mouth. Unwilling to spit out her €50,000 payload, she gamely swallowed and, in obvious pain, attempted to answer Hossein's suitably burning question, 'How did it taste?'

Rützler laughed nervously. 'I was expecting the texture to be more soft,' she said at last. 'There is quite some flavour with the browning. I know there is no fat in it, so I didn't know how juicy it would be, but it's close to meat . . . Er, the consistency is perfect . . . but *I miss salt and pepper*!' With this final outburst, Rützler passed the tasting baton to Schonwald, who soon brought his native burger-eating heritage to bear. 'The bite feels like a conventional hamburger,' he began, 'but it's a kind of unnatural experience, in that I can't tell you how often over the past twenty years I've had a hamburger without ketchup, or any kind of onions or jalapeños or bacon; but I think fat is a big part of what is missing . . . what was conspicuously different was flavour.'

Despite these mixed reviews, Post remained upbeat when asked by Hossein how he felt the tasting had gone. 'I think it's a very good start,' he said; 'this was mostly to prove we can do it. I'm very happy with it. It's a fair comment, that there is no fat in here yet, but we're working on it.' *We*, it emerged, included Google co-founder Sergey Brin, who now appeared in a short film to explain his hopes for the project. 'Sometimes a new technology comes along that has the capability to transform how we view the world,' he said. 'I like to look at the opportunity and see when it's on the cusp of viability.' Brin's speech might have been more uplifting had he not chosen to deliver it in his prototype Google smart glasses, which gave him the sinister appearance of a Bond villain. 'Some people think this is science fiction,' he went on. 'I actually think that's a good thing. If what you're doing is not seen by some people as science fiction, it's probably not transformative enough.'

Back in 2013, Brin was far from the only Silicon Valley CEO getting excited by lab food. That year was something of an *annus mirabilis* for the new tech trend, with Bill Gates announcing his support for no fewer than three start-ups: Nu-Tek Salt, which proposed to replace edible sodium with potassium chloride, Hampton Creek Foods (now renamed JUST), pioneers of the use of plant proteins to mimic eggs, and Beyond Meat, which did the same for chicken and beef. Gates' conversion had apparently come when he tried the latter's 'chicken-free strips' and found that he couldn't tell them from the real thing. 'We're just at the beginning of enormous innovation in this space,' he wrote on his website that year. 'For a world full of people who would benefit from getting a nutritious, protein-rich diet, this makes me very optimistic.'

As usual, Gates was right on the money. Today lab food is big business, with major players including Kleiner Perkins (key investors in Amazon and Google), Vinod Khosla (co-founder of Sun Microsystems) and Obvious Corp (set up by the founders of Twitter) all scrambling for a piece of the action. In just a few years, the science fiction has become reality, with the likes of JUST, Beyond Meat and Impossible Foods (the latter funded by Google, Khosla and Gates) all hitting high-end supermarkets and trendy restaurants in the US and elsewhere. In 2018, the UK got its first taste of 'bleeding' veggie burgers when those made by

Beyond Meat – which, like Post's patty, use beetroot juice for their faux blood – went on sale in Tesco, selling out almost as soon as they touched the shelves. Meanwhile, back in the US reviewers raved about Impossible Food's even gorier fake-blood burgers, which use genetically modified yeast to produce heme (or haem), the compound that gives haemoglobin its name and makes our own blood red. In 2019, Impossible's patties hit the big time when Burger King launched its Impossible Whopper, 'flame grilled to perfection' just like its cow-based counterpart.

With sales of lab meat in the US already at $1.5 billion and projected to rise to $10 billion by 2023, the meat industry has been swift to react. While some producers have demanded that plant-based substitutes not be labelled meat (a rule that the State of Missouri became the first to pass into law in 2018), others, including major companies such as Cargill and Tyson, have taken the 'If you can't beat 'em, join 'em' approach, bankrolling start-ups such as Memphis Meats, which aims to grow in-vitro meat commercially, Post-style, in a lab.

What explains fake meat's meteoric rise? Apart from serious injections of cash, it has been largely fuelled by rapidly rising public awareness of the catastrophic effects of industrial livestock production. From the United Nation's ground-breaking 2006 report *Livestock's Long Shadow* to popular films like *Cowspiracy* and books including Jonathan Safran Foer's *Eating Animals* and recent reports such as the EAT-Lancet Commission's 2019 *Food in the Anthropocene*, a slew of increasingly alarming books, films and studies have emerged documenting the damage, cruelty and ecological lunacy of factory farming.[3]

Humans originally domesticated farm animals largely because the beasts could eat what we couldn't: cows and sheep happily grazed on grass while pigs and chickens gobbled kitchen scraps; after a few years spent in fields, on hills and in backyards – during which the bovines and hens provided us with the added bonus of milk and eggs – we could eat them. Provided one was comfortable with the inevitable endgame, it all created a beautiful, synergetic loop. Factory farming, by contrast, is almost comically inefficient. One third of the global grain harvest is now fed to animals, food which, if we ate it directly, could feed up to ten times as many people.[4] Industrial meat production guzzles one third of all the water used in agriculture and is responsible for an estimated

14.5 per cent of all greenhouse gas emissions.[5] Add in the pollution from the football-pitch-sized pools of toxic slurry and the indiscriminate use of antibiotics and you've got a hefty pile of hidden costs. Although the negative value of such damage is hard to estimate, one study by the Indian Centre for Science and the Environment reckoned that, if you factored everything in, the true cost of an industrial burger would be in the region of $200, not the $2 we usually pay.[6]

The ethical downsides of industrial livestock production are just as troubling. If the term 'factory farm' doesn't immediately arouse a sense of Orwellian disquiet, closer examination of these secretive facilities (known in the trade as concentrated animal feeding operations, or CAFOs) soon will. They are places in which tens of thousands of animals are crowded together and fed on grain and soy-based animal feed aimed at bringing them to slaughter weight as fast as possible. Most of us know deep in our souls that conditions on such farms are far from idyllic. Yet, as Jonathan Safran Foer argues, the price we are prepared to pay for meat directly affects the quality of life that the birds and beasts we eat enjoy, a price that, given Foer's nightmarish travels around various animal gulags in the United States, tends towards rock bottom. As one activist summed it up: 'These factory farmers calculate how close they can keep these animals to death without killing them. That's the business model.'[7]

If part of you still clings to the hope that most of the animals we eat lead happy lives, think again. Of the 70 billion livestock that ended up on our collective global plates in 2018, two thirds were factory farmed; in the USA, the figure was 99 per cent.[8] In order to get some idea of the staggering scale of this, one need only muse on the fact that, of all mammals on earth, 60 per cent are now farm animals, 36 per cent are human and the rest (just 4 per cent) are wild.[9] As such figures suggest, our carnivorous bent is threatening us and our planet.

This is precisely the crisis that Silicon Valley lab-food companies are trying to address. Josh Tetrick, the youthful CEO of JUST, is a committed vegan who began trying to replicate what he calls the 'twenty-two functionalities of an egg' (emulsifying, foaming, thickening and so on) after he discovered the appalling conditions under which most of America's 300 million egg-laying hens were kept. Realising that people weren't going to stop eating eggs any time soon, he wanted, in his words, to 'take the hen out of the equation'. In 2013 he launched Beyond Eggs, a plant-based egg substitute made from ingredients such as peas and oilseed with a longer shelf life, less cholesterol and, Tetrick hoped, a better taste than anything a hen could produce. That same year, Just Mayo went on sale in Whole Foods Market to positive reviews, followed in 2018 by Just Egg, a mung-bean-based substance that resembles scrambled egg when cooked and, if not quite beating the hens at their own game yet, is another step towards Tetrick's vegan dream.[10]

Impossible Foods, Beyond Meat, Beyond Eggs – the language of lab food has a curious air: one of adventure tinged with menace, a bit like some comic-book superhero had accidentally wandered onto the pages of *Brave New World*. The Silicon Valley culture from which it springs – warp-speed technology, megaton profits, ruthless competition and testosterone-fuelled CEOs on a mission to save the planet – has in the space of less than a generation come to dominate most aspects of our lives. So powerful and ubiquitous are the tech giants that it can be a shock to realise just how young they are: in lab food's breakthrough year of 2013, Google was just fifteen years old and Twitter and Facebook less than ten. Despite their youth, such companies wield extraordinary influence, not just over our shopping, communications and personal data, but over how we're likely to live in the future. With profits exceeding $100 billion in 2018, Google had plenty to invest in its newly branded research arm Google AI, leading design and development in every conceivable sphere of our digital existence from face recognition to driverless cars.[11] In retrospect, the question is not what got Silicon Valley into food, but what took it so long.

Whether you find the tech giants' recent obsession with food a cause for celebration or concern will partly depend on your general outlook on life. If you tend to believe that we humans are ingenious enough to invent our way out of any fix, then it's probably time to relax, grab a lab burger and buy shares in JUST and Google. If, on the other hand, you worry that we tend to respond to complex problems by seeking simple solutions, it's time to get very worried indeed. Never before have our lives been more complex, and never before have we been more dependent on technology to make them run. Our planet is in trouble, and a handful of global corporations want to control everything from the way we communicate, travel and inform ourselves to the way we eat. What could possibly go wrong?

As you've probably guessed, I belong to the latter camp. Although I'm no technophobe, I think we urgently need to review our relationship with technology, something I propose in this book that we can do through the lens of food. We humans learned to eat long before we invented technology, and many of our greatest advances have come through our attempts to feed ourselves better. Food and technology are twin pillars of our evolution, so they can help us understand how we arrived at our current predicament as well as guide us in shaping our future. Without further ado, therefore, let's return to Cedric Price's query at the head of this chapter: if technology is the answer, what was the question?

Meat Wave

'How to live?' is the likely answer. The oldest question on earth and one that underpins the actions of every living being, the problem of how to live is wired into our DNA. At its core is the question of how to eat, an issue fundamental to all living things. Trees, frogs, birds, fish and worms all need food, yet the problem for them is far less complex than it is for us. As conscious beings, we feel that there are 'good' and 'bad' ways to feed ourselves. We may not agree on what these are (indeed, wars have been fought over the question), yet eating, for us, has an inescapably ethical dimension.

At the global scale, the problem of how to eat is one we humans are yet to solve. After wrestling with it for the best part of two and a half million years, we still haven't cracked it. We've made some spectacular breakthroughs – we've fashioned tools, tamed fire, invented farming, harnessed steam, modified genes – yet each advance has brought a new raft of problems in its wake. Today our supermarket shelves may heave with food, yet the system that fills them is in crisis. On our finite, overheating planet, the way we eat has trapped us in a vortex of self-destruction from which there is no easy escape. As Robert Malthus warned in his famously gloomy 1798 *Essay on the Principle of Population*, however much food we produce, we always seem to need more.

To compound matters, we're not great at managing the food that we do produce. According to the United Nations Food and Agriculture Organization (FAO), farmers worldwide currently provide the daily equivalent of 2,800 calories of food per person – more than enough to go round, given an ideal food system.[12] Of course, no such system exists, which is why some 850 million people worldwide live in hunger and more than double that number are overweight or obese.[13] The causes of this imbalance are numerous and complex, yet in essence they boil down to the same ones that have dogged our efforts to feed ourselves from the start, namely aspects of geography, climate, ownership, trade, distribution, culture and waste – the very factors that have shaped our civilisations. The way we eat is inextricably linked to the social, political, economic and physical structures that govern our lives, which is what gives food its unparalleled complexity and potential.

Within the complexity, however, certain trends emerge. Developing nations, for example, struggle to produce enough food to feed their citizens, while developed ones tend to overfeed their people. Food waste is a global issue, yet its causes vary depending on where you are: waste in the Global South is mostly due to a lack of infrastructure, while in the North it is largely down to oversupply. The UK, France, Belgium and Italy provide 170–190 per cent of their citizens' nutritional needs, while in the US a gut-busting 3,800 calories are available to every man, woman and child, almost double what can be safely consumed.[14] No wonder so many Americans are overweight, or that half of their food is wasted.[15] As Tristram Stuart pointed out in *Waste*, if Western nations limited their food supplies to just 130 per cent of their nutritional needs and developing states could reduce post-harvest losses to levels similar to those in the developed world, one third of the global food supply could be saved, enough to feed the world's hungry twenty-three times over.[16]

To peel another layer off the food crisis onion, the global diet is changing. As people move to cities, traditional rural diets – typically based on grains and vegetables – are being swapped for Western-style ones with lots of meat and processed foods. The FAO predicted in 2005 that global consumption of meat and dairy would double by 2050, a prediction that remains stubbornly on track.[17] Nowhere is the transition more pronounced than in China, where 80 per cent of the population were rural in 1980, yet 53 per cent now live in cities and 70 per cent are projected to do so by 2025.[18] Back in 1982, the average Chinese ate just 13 kilos of meat per year; today the figure is 60 kilos and rising. Although this is only half what the average American eats, it still means that the Chinese today consume one quarter of all the world's meat and twice as much as the burger-lovin' USA.[19]

In the West, it can be hard to grasp how much meat we eat, since the animals that once grazed in our fields have largely disappeared. Strolling through the British countryside, you could be forgiven for thinking that the entire nation has gone vegetarian, so rarely does one see a cow or sheep. Partly thanks to this mental and physical distancing, many of us live in a state of denial when it comes to our furry and feathered friends: we love our cats and dogs, yet condemn millions of chickens and pigs (the latter being just as feeling and intelligent as our canine companions) to lives of misery. Although welfare standards vary across the world (and British farms have some of the highest), few of us check to see whether the contents of our bacon sarnie came from a 'happy' pig.

Why are we so blind to the truth of food? One answer is that it suits us not to have to think about it too much. To live in blissful ignorance of what it takes to sustain life was once the privilege of the rich; now, thanks to cheap convenience food, most of us can do it. While some might argue that such blitheness is the crowning achievement of industrialisation, it is also symptomatic of a deep moral malaise. Only a scandal on the scale of 'Horsegate' (when cheap meat pies in Europe and elsewhere turned out to contain illegal horsemeat) is enough to wake us from our gastronomic slumber. Immediately after the scandal, British independent butchers reported sales up 30 per cent, as people abandoned cheap pies and sought better alternatives. The renaissance didn't last long, however: sales of pies were back to normal within months, leaving only traces of the crisis in the form of typically British humour. *Waiter*: 'Would you like anything on your burger, sir? *Customer*: 'Yes please, a fiver each way.'

Our dedication to meat-eating is precisely what Mark Post and others hope to address with their lab-grown alternatives. For Post, the advantages of cultured beef are clear: 'It tastes the same, it has the same quality, it has the same price or is even cheaper; so what are you going to choose?' he asks. 'From an ethical point of view, it has only benefits.'[20] Admirable though Post's intentions are, the ethics of lab meat are rather murkier than he makes out. To begin with, cultured beef is grown in bovine fetal serum so, unlike plant-based haem, it still involves the use of animals, albeit on a far smaller scale than conventional beef. Next, there is the squeamish element: the question of whether growing edible muscle tissue in a lab is really an avenue we want to pursue. Last but not least is the problem of ownership: despite Google's unofficial slogan 'Don't be evil' (now modified to 'Do the right thing'), do we really want our food to be made and owned by the same global corporation that controls how we access and share information? And if we don't, who else do we imagine might own the technologies required to make in-vitro beef? Not your friendly local farmer or butcher, that's for sure. If lab meat succeeds – and all the signs are that it will – it is sure to be patented up to its eyeballs and to make profits at least as eye-watering as the software on your smartphone.

So what about plant-based meat alternatives made by Beyond Meat, Impossible Foods and the rest? While less obviously questionable on the ethical front, the jury is still out on whether the mass consumption of such products would really be good for us or the planet. According to its own website, Impossible's Whopper contains: 'Water, Soy Protein Concentrate, Coconut Oil, Sunflower Oil, Natural Flavors, Potato Protein, Methylcellulose, Yeast Extract, Cultured Dextrose, Food Starch Modified, Soy Leghemoglobin, Salt, Soy Protein Isolate, Mixed Tocopherols (Vitamin E), Zinc Gluconate, Thiamine Hydrochloride (Vitamin B1), Sodium Ascorbate (Vitamin C), Niacin, Pyridoxine Hydrochloride (Vitamin B6), Riboflavin (Vitamin B2), Vitamin B12'. Hardly a list of ingredients that Granny would have recognised, let alone trusted.

None of this is to say that lab meat or fake meat is necessarily all bad; on the contrary, anything that promises to end factory farming has to be worth a shot. The problem, as Robert Oppenheimer realised, is that what seems like a good idea in a lab can have unforeseen consequences in the real world. Like dogs, technologies tend to obey their masters, and the behaviour of our current tech giants doesn't exactly inspire confidence in their fitness to control our future food.

The fact that growing muscle tissue in a lab could seem a better idea than simply eating more vegetables reveals the nature of our human dilemma. For millions of years, we've co-evolved with technology, becoming what we choose to call *Homo sapiens* in the process. We wouldn't exist without tech and couldn't survive without it, yet now our co-evolution has hit the buffers. In our bid to solve the problem of how to eat, we've complicated the greater one of how to live. When it comes to tackling that, technology is no longer the limiting factor: we already know how to feed the world, warm and cool our houses and cure disease; what we lack is the capacity to put our ideas into effective practice – to collaborate, share and learn from our mistakes. The areas where we most urgently need to invest and invent aren't technological, they're *human*.

A Good Life

> One cannot think well, love well, sleep well, if one has not dined well.
>
> Virginia Woolf[21]

One question that technology can never answer on our behalf is what makes a good life. The question is central to everything we do, since all our choices and actions are effectively made in response to it.[22] When and how we eat, drink, work, think, walk, talk or check our phones are all decisions steered towards some conscious or unconscious idea of the good. Even when we're asleep, our brains churn away at problems that we failed to solve during the day. Our quest for a good life is one that we can never escape.

When we're hungry, thirsty, cold, ill or in danger, the quest becomes a matter of survival. Food, water, warmth, medicine and shelter present themselves as vitally precious 'goods', as they have been much of the time for most humans in history. Those of us living comfortable lives in the West today are thus something of an anomaly: death, for us, is more likely to come from the so-called 'diseases of affluence' – cancer, heart disease, diabetes or dementia – than from war, violence, starvation or plague. By helping us fight death, technology has distanced us from our own mortality, to the extent that the subject has become largely taboo.

Once survival is assured, the question of how to live becomes increasingly complex and abstract. Although still obliquely concerned with survival *(Have we run out of cornflakes?),* our choices tend towards more intangible aims, such as that of finding happiness. Notoriously hard to define, let alone achieve, happiness is the ultimate tease, universally desired yet rarely possessed. Padding around our heated homes surrounded by computers, dishwashers and microwaves while barking commands at Alexa to play our favourite music, the tacit assumption is that we *ought* to be happy, yet for a plethora of reasons – stress at work, money worries or a pervading sense of loneliness – we can often feel the opposite.

As Richard Layard noted in *Happiness,* the relationship between joy and wealth is far from linear. Once we've reached a certain level of comfort – what is needed for basic subsistence – increased riches don't make us any happier. Indeed, Layard found, despite average incomes in the UK, US and Japan having doubled in the fifty years to 2005, levels of happiness had remained constant.[23] Such findings suggest why even those of us lucky enough to have full bellies, snug houses and smart gadgets feel compelled to reach for something more: love, meaning, fulfilment, purpose. Yet the harder we search, the more unreachable such things can appear. Music, art, astronomy, poetry, philosophy and religion are just some of the outcomes of our yearning, along with base-jumping, Xbox, cryptic crosswords, drugs and alcohol.

Humans are complex beasts, so how can we hope to flourish? Among the first to pose this question was Socrates. Famously provocative and ugly as well as charming and witty, Socrates plagued his fellow Athenians with constant questions about the meaning of life, gleefully picking holes in their answers. He pursued this course because he believed that our greatest task as humans was to learn to use our brains. Needless to say, his efforts didn't go down too well among the ruling elite, who eventually put him on trial for 'corrupting the minds of youth'. In a famous speech, Socrates defended his actions by saying that his greatest insight after a lifetime of enquiry was to realise that he knew nothing. Even so, he said, such questioning was everyone's duty, since a 'life without this sort of examination is not worth living'.[24]

Socrates' devotion to philosophy cost him his life, yet his ideas would prove far harder to snuff out. His relentless search for the meaning of good, immortalised in the *Dialogues* of his faithful pupil Plato, spread throughout Athens, which as the world's first democracy made an ideal context in which to conduct such an enquiry. The real-life city formed the basis of Plato's utopian work *Republic*, which in turn inspired his pupil Aristotle to write his *Ethics*, the first practical guide to leading a good life.

Aristotle agreed with Plato that the guiding principle of life was a search for the good. 'Every art and every investigation, and similarly every action and pursuit, is considered to aim at some good,' he wrote. 'Hence the good has been rightly defined as that at which all things aim.'[25] What then, wondered Aristotle, was the ultimate good for a human? Surely it must mean perfecting our greatest faculty, namely reason? Only through reason, said Aristotle, could we lead virtuous – and thus happy – lives, using it to help us navigate the various pitfalls that life would inevitably throw at us. The key to this was to find a balance in all things, starting with ourselves: if we were naturally impetuous, for example, we should seek patience; if timid, we should try to be braver. Through such efforts, we could perfect our souls and thus achieve virtue, steering a straight path through life much as Odysseus had done when sailing between the monstrous rock and whirlpool of Scylla and Charybdis.[26] If humans were ships and life the sea, then the good was our lodestar and reason our rudder.

No Greek philosopher ever pretended that leading a good life was easy. On the contrary, the idea that it took much courage and effort was axiomatic. The Greek word for happiness, *eudaimonia*, translates as something like flourishing; it was an active, not a passive, state. For Aristotle, this was especially important, since humans were 'political animals', which meant we could never thrive in isolation: in order to be happy, we needed one another. Whatever was good for us had to be good for society as a whole. We may not agree on what that meant, said Aristotle, yet we must still try to find common ground – this, indeed, was the ultimate aim of politics.

As Socrates' fate suggests, not all Athenians felt at ease with such ideas. Yet if leading a virtuous life was hard in ancient Athens, try doing it in contemporary London. In a post-industrial society it is virtually impossible to lead a truly good life, since, merely by existing, we participate in a host of social, political and economic systems that, among other things, oppress workers, abuse animals, poison oceans, destroy ecosystems and churn out greenhouse gases like there's no tomorrow. Heaven help you if you drive a car, fly on holiday, eat steak or own a smartphone. Almost every move we make in the modern world has some distant, negative impact. Just engaging with life's multiple dilemmas requires vast knowledge and effort, as we examine all the implications of our actions on countless people, creatures, structures and organisms, most of which we barely know exist. Needless to say, few of us are equipped for such a task.

What advice might Socrates have given us to help us cope with modern life? His first suggestion might be that we learn to love paradox. Our lifelong pursuit of elusive goals is, after all, pretty paradoxical. Acceptance of the human condition was, for Socrates, the basis of a good life, as it was for his Indian near-contemporary Gautama Buddha. Both men were founders of a tradition of humanist thought that holds such acceptance to be the key to happiness. The idea may sound earnest, yet it is the necessary counterpart to that other great strand of humanism – which remains arguably our best defence against life's vicissitudes – humour. In Douglas Adams' 1970s radio series *The Hitchhiker's Guide to the Galaxy*, for example, a computer called Deep Thought is built for the purpose of answering the question 'What is the meaning of Life, the Universe and Everything?' Deep Thought takes 7.5 million years to come up with the response '42'. When accused of giving a meaningless answer, the computer admits as much, but defends itself by saying that the people who programmed it hadn't understood the original question.[27]

After Easter

Our modern lives are beset by paradox. Our technical capacity is mind-blowing, yet we seem unable to match our skills at, say, genetically modifying sheep, landing probes on comets or making robots serve sushi with non-technical challenges such as creating equitable societies, agreeing to disagree on God or coexisting with fish. In psychological terms, we've developed our 'hard' skills at the expense of our 'soft' ones; in metaphorical ones, we've allowed the technical tail to wag the philosophical dog.

Our dilemma is made worse by the fact that our lives are so dominated by technology. Two thirds of us now own a smartphone, a statistic that underlines the global reach of the digital revolution. The Internet has transformed our lives more profoundly and rapidly than anyone (other than media prophet Marshall McLuhan) could have foreseen. Today we live in a Global Village in which Google is the marketplace, Amazon the general store, Facebook the garden fence and Twitter the local gossip. In the blink of an eye, activities that once only took place in towns or cities can be carried out with a flick of the thumb from a desert, ocean or plane.

Nobody knows where our digital lives will lead. Our screen obsession has already changed our social behaviour and the way we think. Now that the giddy thrill of digital life is starting to subside, its dark side is becoming ever clearer, with reports of cybercrime, self-harm sites, Internet trolls, political propaganda, personal surveillance and data mining coming at us from all sides. The broadening of our communicative horizons has come at the partial cost of our freedom; what once appeared as an innocent new public domain has turned out to be anything but. Drowning in data and hooked on clips of cats doing daft things with dustbin lids, we inhabit a highly manipulated, monetised minefield in which our every move is monitored, stored and sold on for profit.[28] Isolated in our personalised digital worlds and unaware of the algorithms messing with our heads, we are losing our capacity to do what Aristotle called the proper human function: think.

Humans are infinitely inventive and adaptable, yet, as Jared Diamond argues in *Collapse*, we're not very good at noticing when we're in trouble. One of the most haunting tales in his catalogue of doomed civilisations is that of Easter Island. First colonised by the Polynesian Rapa Nui between the seventh and twelfth centuries, the island was a thriving community of some 15,000 souls by the seventeenth, with lush vegetation that included some of the tallest palm trees on earth. With the nearest inhabited land mass more than 1,200 miles away, however, the island was extremely remote and lacked any partners with whom to trade. As the islanders cut down more and more trees to make space to farm, for building materials and for the construction of the massive stone heads *(moai)* for which the island is famous, soil erosion reduced their capacity to grow food. More ominously still, the lack of trees meant that they could no longer build boats with which to go fishing. When Europeans finally arrived in 1722, they found a malnourished population of less than 3,000, widespread evidence of fighting and a denuded landscape in which no tree stood taller than ten feet.

Easter Island, Diamond argues, makes an apt metaphor for planet earth. Although rats and disease from visiting ships delivered the *coup de grâce*, what really sealed the islanders' fate was their isolation. Not that we are necessarily doomed to follow the islanders into oblivion: as Diamond notes, of all the reasons why societies collapse (environmental damage, climate change, hostile neighbours or trade partners), 'the fifth set of factors – the society's responses to its environmental problems – always proves significant'.[29]

What might the Rapa Nui be able to tell us about our modern predicament? Like them, we know we're living beyond our means; like them, we're not reacting fast enough. We know that we need to change our ways, yet the complexity of the threats we face seems overwhelming, so we end up carrying on as we are. We urgently need new ways of thinking that don't send our brains into gridlock and that, crucially, deliver a new vision of how we might live in the future. All of which brings us back to food.

Sitopia

> It is gastronomy which makes a study of men and things.
> Jean Anthelme Brillat-Savarin[30]

Food shapes our lives, so it can help us think. We may not be aware of its influence, but it is everywhere: even in the parts of our brain that can't stop wondering about the meaning of life. Food's effects are so ubiquitous that they can be hard to spot, which is why learning to see through the lens of food can be so revelatory. One perceives a remarkable connectivity: an energy that flows through our bodies and world, linking and animating everything as it goes. As we have seen, I call this food-shaped world *sitopia*.[31] Unlike utopia, which is ideal and therefore can't exist, sitopia is very real. Indeed, we already live in it; just not in a very good one, since we don't value the stuff from which it is made.

Sitopia is essentially a way of viewing the world. Food can help us to understand complexity, since it represents life, yet is material and graspable. Whether or not we think about it, we all intuitively comprehend food: Descartes might just as well have said, 'I eat, therefore I am.' This instinct is hugely powerful, since it links us directly to our past: our ancestors lived very different lives to us, yet they too had to eat. People's efforts to feed themselves have shaped every human society and thus represent a vast repertory of ideas, thoughts and practices from which we can draw. We can use the lens of food as a conceptual time machine, to help us view our past and perceive our present, and thus imagine a future in which we know food will remain pivotal.

Before we can use food as a lens, however, we need to learn to see food itself, a task made harder by the fact that, although the act of eating is universal, it is also highly personal. Food culture is a language that we learn so early in life that we don't realise we've learned it. As omnivores, we can adapt to eating almost anything, yet we are not born instinctively knowing what to eat; that is something that we start learning from our very first meal. As newborn infants, we eat before we think; eating thus predates consciousness. From our first gulp of mother's milk to our last supper, meals determine the shape and rhythm of our lives, forging our bodies, tastes, social bonds and identities. As children, we first learn how to eat with family and friends, and by the age of three or four our habits are already becoming ingrained. From now on, our reactions to unfamiliar foods are likely to be more cautious: as we grow older, we may start to find other people's food habits unappetising, puzzling or even repugnant.

On a trip to Thailand some years ago, I had to confront my own food prejudices when I was taken to a jungle market specialising in insects. Many Thais crave insects much as the British do chocolate, but as I contemplated the carpet of shiny critters on offer for my lunch, I could feel my stomach shrinking. Eventually, I summoned up the courage to try a cricket, telling myself that it was really just like a prawn with wings. Putting it into my mouth, I found it to be crunchy, salty, fishy and meaty – in short, perfectly delicious. However, forty years of conditioning won the day: even though I managed to swallow the bug, for days afterwards the thought of it made me feel decidedly queasy.

The uneasiness we often experience when faced with unfamiliar foods is in stark contrast to the comfort of eating familiar ones, especially those from our childhood. The taste of such dishes can create a powerful sense of nostalgia, even if, as the British cook and food writer Nigel Slater notes in his autobiography *Toast*, the food itself wasn't all that tasty. What matters most is that it was made with love. Slater recalls that the icing on his mother's Christmas cake was so hard that even the dog wouldn't eat it, yet 'I believed that cake held the family together. Something about the way my mother put a cake on the table made me feel that all was well. Safe. Secure. Unshakeable.'[32]

Food is so bound up with our sense of self as to be virtually indistinguishable. We've all got food stories, memories, habits and preferences, meals that we love or hate, yet one thing that most of us share (unless we are afflicted by some trauma or illness) is the pleasure of eating itself. As the French 'philosopher of taste' Jean Anthelme Brillat-Savarin noted in his 1825 treatise *La Physiologie du Goût*, eating is our most reliable and enduring joy:'The pleasures of the table belong to all times and all ages, to every country and every day; they go hand in hand with all our other pleasures, outlast them, and remain to console us for their loss.'[33]

Food culture goes to our very core. How we produce, trade, cook, eat, waste and value food says more about us than we realise: such practices form the structures upon which our lives are built. Food is both the substance of life and its deepest metaphor.

The Endless Meal

Contemplating food is something we rarely do in the modern world; industrialisation has done its best to obscure the origins of what we eat. Thinking about what food really is can make us feel uneasy, since it brings us uncomfortably close to examining the nature of our own being. This was, however, the precise realisation that propelled Charles Darwin towards his greatest discovery. Struggling to explain the great variety of species on earth, it dawned on Darwin that competition for limited resources meant that only those creatures most suited to a particular environment would survive to reproduce. 'Survival of the fittest' would lead to specialisation that over time would evolve into a profusion of different species.

Darwin's thinking led him to a disturbing conclusion: stripped of table manners, man's need for food was no different to that of any other creature. All species, including humans, he realised, were in competition for the same resources. The need to eat thus joined all living beings in a perpetual mutual carnage, such as that which underpinned even the most seemingly innocent springtime scene:

> We behold the face of nature bright with gladness, we often see superabundance of food; we do not see, or we forget that the birds which are idly singing round us mostly live on insects or seeds, and are thus constantly destroying life; or we forget how largely these songsters, or their eggs, or their nestlings, are destroyed by birds and beasts of prey; we do not always bear in mind, that though food may now be superabundant, it is not so at all seasons of each recurring year.[34]

The fact that Darwin was reading Malthus's *Essay on the Principle of Population* when the penny dropped was no coincidence. Malthus had argued that populations were limited by the amount of food available, which meant that, since populations grew geometrically while food supplies increased only arithmetically, humans were doomed to eventually run out of food. If society were to avoid the misery of 'positive checks' on population growth (hunger, disease and war), it would have to reduce the number of people through the exercise of 'moral restraint', aka birth control. Malthus's theories proved instantly controversial and have remained so ever since, yet they provided Darwin with a key part of his evolutionary puzzle. To the idea that humans and animals had evolved from common ancestors he added the notion that the struggle for life was in effect a never-ending meal in which all living beings took part.

The 1859 publication of *On the Origin of Species* lobbed a grenade into the heart of natural science. The idea that men were somehow physically and genetically bound to their fellow creatures was one that most Victorians found hard to stomach. Yet, despite its rocky reception, Darwin's theory shifted the axis of thought concerning man's relationship with nature, with insights that remain powerful even today. Whenever we earthly creatures eat, we eat together; furthermore, we eat one another. Food consists of living things that we kill in order to live – provided we have the means and desire to eat them.

Humans don't only kill for food, of course; alone among species, we also grow and breed the things we eat. This makes the question of feeding ourselves doubly tricky. A lion doesn't lie awake at night worrying about whether or not it should have eaten that baby gazelle; for a lion, how to eat is a practical, not a moral problem. Yet, as Darwin noted, the effects of the lion's decisions over time will nevertheless teach it how to eat: if it guzzles too many gazelles, it will eventually run out of food. When it comes to keeping a balance between eaters and eaten, nature has ways of maintaining the status quo.

For us, however, it's a different story. Thanks to modern agriculture and medicine, we've dodged the famines and pandemics that Malthus predicted would naturally curb our numbers, leading to an unprecedented population explosion. During the twentieth century alone, our numbers rose from 1.7 billion to over 6 billion, a growth made possible in large part by the 1909 discovery by the German chemist Fritz Haber of how to 'fix' atmospheric nitrogen (i.e. turn it into the compound ammonia) and so make it available to plants. Artificial nitrogen is the key ingredient in chemical fertilisers commonly known as NPK, so named since they also contain phosphorus (P) and potassium (K).[35] Today, the so-called Haber–Bosch process (Carl Bosch industrialised

Haber's idea) is credited with feeding an estimated two out of every five people on the planet.[36]

Malthus's critics argue that such technical breakthroughs blow his theory sky-high. Had he lived to see modern agriculture, they claim, he would have realised the error of his ways. Malthus was just a misanthrope who failed to see that human ingenuity will always triumph. Malthusians counter this argument by pointing out that, while the Haber–Bosch process is undoubtedly clever, the practice of bombarding the soil with chemicals doesn't do it any good in the long run. Indeed, by providing the means by which the population has grown exponentially, it merely ramps up the pressure on other natural resources upon which we also rely. Man cannot live on NPK alone.

Having had the temerity to mention food, death and morality all in the same breath, Malthus is, perhaps inevitably, the figure around which the 'feed the world' debate tends to galvanise. By raising the issue of population, he ventured into territory that for many remains taboo even today. Yet to discuss how we should eat *without* addressing the question of population is at best limited and at worst meaningless, since the two problems are so obviously connected. Malthus may have been a doom-mongering pessimist, but his theory is yet to be proven wrong. However responsibly we farm, fish, hunt or gather, our appetites continue to shape the planet and affect the life chances of us and our fellow earthlings.

Adam's Apple

In order to live we must eat; in order to eat, we must take life. This circularity may seem remote when most of our food comes ready-cooked in boxes, yet its logic underpins our very existence. Whenever we eat, we make an implicit value judgement: that human life is worth more than that of, say, a leek. Most of us agree on this: it is, after all, the basis upon which we all, including vegans, sustain ourselves. But what of lamb? Vegetarians draw the line here, although they still eat eggs and cheese, the production of which also involves the taking of animal life. Carnivores eat lamb, although conscientious ones will insist that the animal has had a good life and (so far as such a thing is possible) a good death first.

Such thoughts are hardly appetising, so it is fortunate that we don't have to construct the ethical universe every time we want to eat breakfast, any more than we have to work out how to start a fire or make toast. Our ancestors did all the hard work for us, both by learning which plants and animals were edible or poisonous (not a job high on the health and safety index) and by creating the framework within which we still eat today. The complex array of rules, habits, skills, knowledge, dos and don'ts that they've handed down to us is what we call food culture.

Our ideas about food are shaped by this cultural frame to such an extent that only *in extremis* (and sometimes not even then) can our concept of what is or is not edible be shaken. A notorious problem with food aid, for example, is that people used to eating certain foods such as yams might refuse well-meant gifts of wheat, even to the point of starvation. Wheat, to them, is simply not food. Occasionally, however, a crisis can override even the strictest norms, as was the case during Sir John Franklin's doomed 1845 expedition to find the North-West Passage. Trapped in the ice, the crew became so desperate that some resorted to cannibalism, yet seemed unable to seek help from the local Inuit, who lived perfectly well in those icy climes. The Inuit reported seeing starving sailors staggering past in an attempt to escape their plight without thinking to stop and ask them for aid.

Even in secular societies, our ideas of what can and cannot be eaten are powerfully reinforced by myth. In the Judeo-Christian world, for example, the rules are laid out in the very first chapter of Genesis, in which God establishes man's dominion over his fellow creatures by telling Adam, 'See, I have given you every plant yielding seed that is upon the face of all the earth, and every tree with seed in its fruit; you shall have them for food.'[37] Adam is thus created vegan, and his home, the Garden of Eden, is a fruitarian paradise in which he and Eve are free to wander at will. There is, however, a catch: 'You may freely eat of every tree of the garden,' warns God, 'but of the tree of the knowledge of good and evil you shall not eat, for in the day that you eat of it, you shall die.'[38]

Eve can't resist the forbidden fruit, of course, leading to the Fall of Man which marks the start of the human story in the Bible. Cast out of Eden, Adam and Eve are condemned to farming: a way of life considered far harder than hunting and gathering in the ancient world. As a result, they become omnivores, a transition that brings its own burden of woe for their sons, when God prefers Abel's sacrifice of lamb to Cain's offering of grain, leading to a fit of jealousy in which Cain murders his brother.

Food features heavily in Genesis for good reason. The narrative mirrors our human journey from hunter-gatherer to farmer and citizen, documenting the struggles and sacrifices that must be made along the way. As life gets progressively more complex, the protagonists wrestle with a succession of Aristotelian dilemmas: between innocence and knowledge, freedom and obedience, power and responsibility. Life is presented as a series of tests which people usually fail, yet in the process of facing such challenges, they become ever more human. Crucially, the whole journey begins with the dawning awareness of good and evil: living with such knowledge, the story suggests, is humanity's unique burden.

With the exception of a few dietary quibbles, the Old Testament establishes the right of humans to be omnivores, an assumption that still prevails in the West. Other traditions, however, have viewed things very differently. In India, for example, vegetarianism has long been customary, since Buddhists, Jains and Brahmins all reject animal slaughter, while Hindus avoid both beef and pork, and Muslims eschew the latter. India's sacred cows are a striking symbol of this different belief system: revered for their life-giving milk, the animals wander freely and are fed by people as they pass. As Reay Tannahill explained in *Food in History*, the origins of such food cultures are often to do with practicality (cows were rare on the Indian subcontinent and could feed many more people if kept alive). Yet they also reflect very different views of life itself: in the Hindu Vedas, for example, all living things have souls that are endlessly reincarnated depending on their karma, a spiritual force influenced by earthly acts, especially violent ones.[39]

Such a world view naturally makes the business of eating rather tricky. Jains, for example, practice *ahimsa*, or non-violence, so meat, fish, eggs and dairy are off the menu. The eating of honey is forbidden, since it is considered harmful to bees, and strict adherents also avoid tubers, on the basis that microorganisms might be harmed when digging them up, which cuts out potatoes, onions and garlic. At its most ascetic, the Jain diet thus consists only of fruit and vegetables, nuts, pulses and grains, permitted on the basis that no human can exist without taking at least *some* life.

The contrasting world views of East and West are reflected, as one might expect, in similarly divergent attitudes towards mortality. While we in the West try to prolong our lives at any cost, a Jain who feels that she has accomplished all that she can in this life may perform the highly respected act of *santhara*, a deliberate fasting to death. To a Westerner, such a deed might seem puzzling or tragic; to a Jain, our tendency to cling to life would seem equally bizarre. Our world views are shaped by the cultures into which we are born, yet whatever our outlook on life, our common need to eat transcends them all.

The Modern Epicure

> Nothing is sufficient for the person who finds sufficiency too little.
>
> Epicurus[40]

For the Greek philosopher Epicurus, satisfying one's appetite was central to living well. In his garden overlooking Athens, Epicurus welcomed men and women from all walks of life, including slaves, to share a simple meal of home-grown vegetables, bread and water with him, perhaps with some cheese and wine, to be eaten while discussing life, the universe and everything. Learning to savour such simple pleasures, Epicurus believed, was the key to happiness. Rarely has an idea been so widely misrepresented: today, an epicure is synonymous with a gourmand, someone whose refined tastes, knowledge and wallet allow them to appreciate the very finest that haute cuisine can offer. To Epicurus, however, such sophistication was a guaranteed road to ruin:

When I say that pleasure is the goal of living, I do not mean the pleasures of libertines or the pleasures inherent in positive enjoyment . . . I mean, on the contrary, the pleasure that consists in freedom from pain and mental agitation. The pleasant life is not the product of one drinking party after another or of sexual intercourse with women and boys or of sea food and other delicacies afforded by the luxurious table.[41]

If you read these lines, as I did, with a twinge of disappointment, you may struggle to live according to the strictest Epicurean principles. There is, however, more method – and joy – to Epicurus' asceticism than meets the eye. Like animals, said Epicurus, we find pleasure in simple acts, such as satisfying our hunger and thirst. When we enjoy a cool drink after a long hot walk, for example, we feel a surge of pleasure as our thirst is quenched. We experience such pleasures as a natural good, which is why humans, like all other animals, are natural hedonists. Since we instinctively register such pleasures as good and pains such as hunger or thirst as bad, such sensations have inherent value for us.

Most of us can probably go along with Epicurus thus far – few of us, after all, need much prompting when it comes to seeking pleasure. But here is the rub: for Epicurus, the pleasure to be gained from satisfying one's appetite with a simple meal of bread and water was *as good as it gets*. Any attempt to crank up the hedonometer by adding some tangy goat's cheese or fragrant wine did not increase the pleasure of eating, said Epicurus, but merely altered its nature. Furthermore, the future pain that such indulgences might bring through, for example, biliousness or a raging hangover, further reduced their overall benefit. Graver still were the dangers of regular gourmandising, which left one constantly craving delicacies and reduced one's ability to enjoy plainer fare. Far better, said Epicurus, to learn to savour everyday pleasures than to crave treats that one could obtain only rarely.

If you are beginning to think that Epicurus' philosophy has something of an Eastern flavour about it, you would be right. Among his early influences were the philosophers Democritus and Pyrrho, both of whom travelled to India, where Vedic religions already flourished. It is no accident that the Greek concept of *ataraxia* – freedom from mental anguish – has strong echoes of the Buddhist concept of nirvana, release from suffering. For Epicurus, *ataraxia* represented the highest possible good, one that could only be achieved by banishing irrational fears, such as those of death and the gods. We have nothing to dread from death, he argued, since it is merely non-existence that we can't even experience, while the gods are far too busy with their own affairs to bother with ours. Having dismissed such baseless fears, therefore, we can enjoy a happy life, pondering the meaning of life in the company of friends.

Few of us have the strength to live fearlessly, let alone to subsist on bread and water. Yet in identifying the pleasures to be gained from simple things – by finding joy in the everyday – Epicurus had an insight remarkably in tune with contemporary psychology. The pleasure principle that he identified is increasingly recognised as key to personal motivation. Similarly, his recognition that we must seek happiness beyond materialism reads just like a modern mindfulness manual: 'Nothing is enough for the man for whom enough is too little,' he declared. But then, he had never seen an iPad.

Food for Thought

What would a time-travelling Epicurus make of our world today? Our consumerism would no doubt horrify him, as would our habit of gorging on fast food. But his greatest concern would probably be in discovering how little time we spend in reflection. A modern Epicurus would no doubt find his place in the blogosphere, yet might struggle to deal with the complexities of modern life. Even in his own day, Epicurus had little truck with politics: in contrast to Socrates, who loved the bustle of the agora and lived and died by the laws of Athens, Epicurus retreated to the sanctuary of his garden. Some critics have labelled this withdrawal naïve or selfish; yet Epicurus' focus on individual virtue perhaps explains why he speaks to us so directly today, in an era when personal fulfilment is paramount and identity politics holds sway. Yet, as Aristotle noted, the distinction between private and public flourishing is ultimately false: no matter whether one engages with life's dilemmas on the social or the individual level, the only true answers must balance both.

The thinker who reframed the Greek concept of virtue for the modern age most directly was arguably the American psychologist Abraham Maslow. In his 1962 book *Towards a Psychology of Being*, Maslow argued that all humans have a hierarchy of needs, from physiological ones (for food, water and sleep) to safety (shelter and peace), love (family and belonging), esteem (status and recognition) and finally self-actualisation (expressing one's inner nature). Maslow's hierarchy of needs has a clear order of priority: if one is starving, one tends to search for food rather than write poetry. This does not mean, however, that our 'higher' needs are any less important than our basic ones, as Maslow makes clear: 'It would not occur to anyone to question the statement that we "need" iodine or vitamin C. I would remind you that the evidence that we "need" love is of exactly the same type.'[42]

Self-actualisation, for Maslow, was the goal of a good life, echoing Aristotle's concept of the perfected soul. Before we could practise it, however, our basic needs had to be met: necessities that Maslow labelled 'deficiency needs', as opposed to the sole 'growth need' of self-actualisation. Also following Aristotle, Maslow acknowledged the crucial role of society in satisfying such necessities: 'The needs for safety, belongingness, love relations and for respect can be satisfied only by other people, i.e. only from outside the person. This means considerable dependence on the environment.'[43]

As children, said Maslow, such dependence is natural, since we must rely on our parents to supply all such wants. If these are not fully met, however, we risk becoming needy adults, constantly seeking reassurance, approbation and affection from others.[44] Even in such a case, however, all is not lost. With the right support, we can still become self-actualisers, a process that is by its very nature healing. All we need do is switch from deficiency-motivated acts to growth-motivated ones, by seeking the 'higher' pleasures of craft, creativity and insight. For Maslow, this inner shift changed everything, since it introduced a kind of engagement that was self-generating:

> With growth-motivated people, gratification breeds increased rather than decreased motivation, heightened rather than lessened excitement. The appetites become intensified and heightened. They grow upon themselves and instead of wanting less and less, such a person wants more and more of, for instance, education. The appetite for growth is whetted rather than allayed by gratification. Growth is, *in itself*, a rewarding and exciting process.[45]

Most of us are probably familiar with experiences such as those which Maslow describes: learning to play a musical instrument, perhaps, to cook delicious food or move it like Messi on the football pitch. Engaging in skilled activities like these has been described by the psychologist Mihaly Csikszentmihalyi as optimal experience or 'flow'.[46] Such focussed practice is a natural win-win, he says, since the more one does of it, the more pleasure one derives. In contrast to tasks that we perform for money, striving for accomplishment becomes an end in itself. We crave it like an addiction, yet unlike drugs, which dull our faculties, it rather sharpens them. To the ancient Greeks, for whom gymnastics were a natural adjunct to thinking, performing such skills would have been well understood as a way to cultivate virtue.

Because they depend on our inner development, the pursuit of self-actualisation and flow counterbalance the consumption-based growth that is theoretically supposed to deliver happiness in the modern world. A great violinist doesn't throw away her violin every year to buy a new one, but rather hones her skills on a trusted instrument that she will keep for life. Similarly, a good farmer doesn't destroy his soil, but rather builds up its fertility over time. If we were to shift away from consumerism towards cultivation – from what one might call external to inner growth – it would have far-reaching consequences for our way of life and for the values underpinning our economy.

All of which is why, in our modern quest for virtue, food occupies a unique place. As the one thing we must consume every day, our most reliable source of pleasure and the cause of our greatest demands on the natural world, it embodies more directly than anything else the battle between our internal and external needs. Learning to value it and to see through it thus represents our best chance of balancing the two.

2
Body

Weighing In

'OK, let's do this!' Pam, a fit-looking thirty-something blonde, smiles at me encouragingly. I take a deep breath and climb onto the scales. As usual, the result is somewhat dispiriting. 'Is that what you were expecting?' asks Pam, a note of sympathy in her voice. Actually, it was. Like most people who love their food, I've got a fair idea of what I weigh, and it's usually too much. Along with millions of others, I've spent much of my life on a variety of diets. Atkins, Dukan, 5:2: I've tried them all, and most have worked, at least for a while. But like waves on a rising tide, the pounds have always crept inexorably back. The fact that I'm fond of bread, butter, cheese, chocolate, potatoes, pasta and wine – indeed, pretty much all food – doesn't help. Nor does the fact that I spend most of my life sitting at my desk. There is, however, one diet that I've never tried, largely because the idea of being weighed in public and paying for the privilege has never appealed to me. But since I've exhausted all other options, I now find myself at my first Weight Watchers weigh-in.

'This is all about making food work for you!' says Pam with the eagerness of a convert. 'You can eat all your favourite foods and tailor the diet to suit you; just plan your meals ahead and track what you eat. *Good things are coming!*' Based on my current weight, Pam allocates me 29 daily 'food points' that I can 'spend' in any way I like, with an additional 49 for treats throughout the week. She hands me a small booklet, not unlike a ration book, in which hundreds of foods are scored according to their relative sinfulness. I glance down the list to see what this is going to do to my regime. Most fruit and vegetables are 'free', which is good, but other items mount up pretty fast: a skinless chicken breast is 4 points and a small glass of wine ditto, while a 40-gram piece of cheddar (the miserly match-box sized amount one used to get served on a plane) is 5. My favourite late-evening snack of cheese and biscuits is clearly going to have to go, but I guess I already knew that.

Pam shows me a picture of herself taken three years ago, when she was three stone heavier. She looks smiley and chunky in the picture, much as I did during the chubbier phases of my fourth decade. I tell her I am impressed by the transformation. 'Yes, that was me,' she says in the sort of wistful tone one might use to describe a long-lost family pet. 'Rather different to now.' The fact that Pam has maintained her 'goal weight' for several years means that she can now act as a Weight Watchers leader, living proof that this diet can really work. In this, she is following in the footsteps of Weight Watchers' founder, American housewife Jean Nidetch, who in 1961 asked her friends to help her lose weight by coming to weekly weigh-ins at her house. When she successfully shed the pounds, Nidetch had the idea of setting up similar meetings for other women, using her own example to inspire others. Today, 40,000 such meetings are held each week worldwide, attended by one million members.

As I chat with Pam, the room, which is in a Salvation Army building in central London, starts to fill with people coming for their weekly weigh-in. Apart from one American man who just seems to be here for a natter, all are women. They come from all walks of life, young and old and all shapes and sizes, but none of them is vast; I suppose the fact that they are prepared to give up their lunch break in order to be here explains why. Some are regulars whom Pam greets by name; others are lapsed members who must re-register before getting back on the scales. Admin complete, the women start forming an orderly queue, shedding coats and shoes as they approach the moment of gravitational truth, like passengers going through security at an airport.

Presiding at the scales, Pam keeps up a steady stream of chat. 'Hello, nice to see you again. How's it been?' 'Not too good,' says a gloomy young woman, only to be contradicted by the scales. 'What do you mean, not good, you've lost *three pounds!*' Pam shrieks, and the woman comes away blushing with pride. Her successor, however, is less fortunate: 'I can't understand why I haven't lost anything,' she says. 'I've been so good.' 'Never mind,' coos Pam. 'We can get you back on track. Good things are coming.' And so it goes: the queue shuffles forward as each dieter receives their reckoning, with some murmurs of praise or encouragement. I am suddenly reminded of being in church, watching as people come forward to take Communion. Suddenly, there is a shout from the front: '*Seven and a half!*' a woman squeals. 'I was only expecting three!' 'Oh, well done, that's amazing!' says Pam. The woman returns from the scales bubbling with excitement, and for a moment the room is transfigured by her epiphany. *This* is what we've all paid our weekly £6.25 sub for: the transformative joy of waving goodbye to some unwanted avoirdupois. Everyone exchanges smiles.

As people hang about chatting after the weigh-in, I browse the size-able range of Weight Watchers products – mostly biscuits, sweets and chocolate – on display by the till. One middle-aged woman chooses a large packet of biscuits. 'My son eats them all,' she says apologetically. 'He seems to prefer Weight Watchers biscuits to real ones.' I decide to try something too, and go for some chocolate bars filled with caramel, biscuit and 'chewy whip', at a mere 84 calories (and 2 points) a shot. Before I leave, Pam hands me a book full of suggested menus and re-cipes and a personal record card, with my start weight noted down and future weeks of theoretically diminishing body mass stretching off seemingly to infinity.

It's a lovely spring day, and I decide to walk back through the park, earning myself a couple of bonus points on the way. After this morning's session, I'm not sure if this diet is for me: all the counting seems tire-some, and the weekly weigh-in feels a bit too much like going to the gym. Approaching a patch of yellow daffodils, I decide to sit down and try one of my chewy chocolate bars. Peering at the tiny writing on the wrapper detailing its lengthy list of ingredients, I find they include some dodgy-sounding substances such as 'bulking agents', although, on the positive side, they also appear to contain real chocolate. Throwing caution to the wind, I take a bite. Perhaps because it's a beautiful day and I'm full of the joys of spring, the bar tastes surprisingly good. Yet part of me can't help feeling that the very idea of diet chocolate is somehow wrong.

Food and Us

> The pleasure of eating requires, if not hunger, then at least appetite.
>
> Jean Anthelme Brillat-Savarin[1]

My first and probably last Weight Watchers experience gave me a glimpse into a lucrative global business. Today, Weight Watchers (or WW, as it was rebranded in 2018) boasts Oprah Winfrey as a major shareholder and spokesperson and its calorie-controlled convenience foods and 'wellness' programmes generate annual revenues of $1.3 billion.[2] Impressive though that is, it is a mere drop in the flab-busting ocean: in 2019 the US diet industry was worth $72 billion.[3]

Such figures raise big questions about our relationship with food. With 45 million Americans going on a diet each year, the US leads the field as far as obesity and dieting go, yet is by no means alone: one in four Britons is on some form of diet at any one time, and the phenomenon, like our waistlines, is spreading.[4] So why do we spend billions on slimming products when we could just eat a bit less and exercise more? Or, to put it another way, why are we getting so fat, and why do we seem so helpless when it comes to doing anything about it?

One reason is that our responses to food are involuntary. I've had some delicious meals in my time – Michelin starred and all that – yet one of the most enjoyable things I've ever eaten was a bar of Cadbury's Dairy Milk that I demolished at the age of fifteen at the top of a craggy peak in the Lake District. I had just slogged up the mountain in heavy fog and was ravenous, and the chocolate hit the spot. When we're hungry, I learned that day, food is inherently delicious. The reason why is obvious: since we've got to eat in order to stay alive, our bodies reward us for eating: 'delicious' is just body-speak for 'thanks, and keep it coming'. Fortunately, some cooks have the skill to transport us to gustatory heaven without the need to slog up a mountain first. The point is that, when we have an appetite, common-or-garden food can do the same. Epicurus was spot on.

But few of us in the West these days wait until we're hungry before we eat. Once frowned upon, snacking between meals is now the norm: Americans eat on average five meals or snacks a day, with just one quarter following the traditional pattern of breakfast, lunch and dinner.[5] In the

UK, 57 per cent admit to skipping meals and snacking instead, with 30 per cent doing so at least once a day.[6] Since our jobs are increasingly sedentary, we don't work up as much appetite as our more active ancestors either. As a result, we often eat more out of habit than hunger, and increasingly while doing something else. On a recent trip to Chicago, I was amazed to see that my taxi driver had a plate of pasta on his lap, which he proceeded to scoop up in dollops with one hand while steering, erratically, with the other. Although alarming to the uninitiated, such sights are increasingly commonplace in the US, where one in five meals is now eaten in a car.[7]

As the world's leading industrial food nation, the United States is the place to which we must turn in order to understand how our relationship with food is changing. America pioneered most of the products and processes – factory farming, flash freezing, supermarkets, fast food – that now dominate the global food system. As a consequence, it has long been the fattest nation on earth.[8] Travelling around the country, it's not hard to see why: streets, malls, parks and museums overflow with eateries, and portion sizes are humungous – one sandwich I bought at O'Hare Airport would easily have fed a family of four. In the US and increasingly elsewhere, mealtimes have morphed into one seamless eating opportunity. It's become normal to wander around clutching a soda or coffee, and seats in cinemas now come fitted with trays so that you can eat a full meal while watching your movie. This constant pressure to eat creates what is known as an obesogenic society, a place where simply living there makes you fat.

Today, America is far from the only nation with an obesity problem; anywhere that adopts its food culture – as Mexico and the UK did decades ago – is going the same way.[9] So why do we find it so hard to resist hamburgers, doughnuts and pizza? The answer lies in the way our bodies reward us for eating and also our susceptibility to getting overly attached, as Epicurus warned us we might, to pleasure.

A Matter of Taste

It all comes down to taste. Our most universal hedonistic stimulus, taste is the sense through which we sample the world's chemical composition.

At a molecular level, it allows us to distinguish nourishment from poison, i.e. to tell good food from bad. As Harold McGee points out in his 1984 book *On Food and Cooking*, this is a skill that even single-celled organisms need: protozoa, for example, will move towards sources of sugar and avoid poisonous alkaloids. As McGee notes, this makes taste our most primal sense: 'Because nutrition is a matter of finding and ingesting particular chemical compounds, some such sense has been a necessity from the very beginnings of life.'[10]

Most of us can name five basic tastes – salty, bitter, umami, sweet and sour – that we register via our taste buds, the clusters of sensor cells visible on our tongues. When we eat, our saliva creates a solution that allows the taste buds to sample the chemical makeup of the food, sending the resulting data directly to our brain. All five tastes have vital meanings in the wild: sweet, salty and umami foods such as fruit, fish and seaweed are generally good for us, while bitter or sour substances can be poisonous. This natural wisdom of the palate was recognised by Jean-Jacques Rousseau: 'If we had to wait till experience taught us to know and choose fit food for ourselves,' he wrote in his 1762 novel *Émile*, 'we should die of hunger or poison; but a kindly providence which has made pleasure the means of self-preservation to sentient beings teaches us through our palate what is suitable for our stomach.'[11]

Although taste buds determine the basics of flavour, smell is the sense that delivers it in its full technicolour glory. Before we eat, volatiles (airborne molecules) from food are picked up by olfactory cells at the top of our nose, sending signals to the brain telling it that food is on its way – which is why the mere smell of baking can make us drool. As we start to chew, more volatiles travel past our throat and up through our nose, providing us with a second reading. Only when these orthonasal and retronasal signals combine with those from our taste buds do we finally 'taste' food. Tasting is thus an act of triangulation in which we experience flavour, not in our mouth, but in our orbitofrontal cortex, just behind our eyes.[12] The fact that this part of the brain is directly connected to regions responsible for memory and emotion explains why certain tastes can trigger a powerful nostalgia, as is famously the case for the narrator of Marcel Proust's *À la recherche du temps perdu*, for whom the taste of a madeleine dunked in tea brings back such a rush of childhood memories that they go on to fill seven volumes and over 4,000 pages:

> No sooner had the warm liquid, and the crumbs with it, touched my palate, than a shudder ran through my whole body, and I stopped, intent upon the extraordinary changes that were taking place. An exquisite pleasure had invaded my senses, but individual, detached, with no suggestion of its origin. And at once the vicissitudes of life had become indifferent to me, its disasters innocuous, its brevity illusory – this new sensation having had on me the effect which love has, of filling me with a precious essence; or rather this essence was not in me, it was myself.[13]

Our relationship with flavour is particularly personal since, unlike with sights and sounds, we have no way of replicating taste and smell. Although we often share food, our sensory experience of eating remains essentially private. In addition, new research is revealing the highly individual nature of our palates: our sense of flavour varies more widely than our capacity to see or hear. For example, one in four of us is a 'super-taster', with up to sixteen times as many taste buds as 'non-tasters', who also represent one in four. Although super-tasters can distinguish flavours better than others, there is a downside to their talent, since they can find bitter tastes – such as those found in the notorious Brussels sprouts – unpleasant to the point of pain. Since super-taster genes are not necessarily inherited, such differences can make family mealtimes particularly fraught.

Even if we don't inherit our parents' taste buds, our mothers can still influence our preferences before we are born. Pregnant women partial to curry, for example, can pass strong flavours like garlic and chilli through their amniotic fluid to their future offspring, giving the latter a penchant for spicy foods later in life. Most of our taste preferences, however, are acquired after we're born, as any parent who has resorted to the 'train going into the tunnel' routine knows only too well. When it comes to liking food, familiarity is everything, and some foods may have to be tasted sixteen times before children will accept them. As Bee Wilson pointed out in *First Bite*, the temptation to subvert this process by bribing children to 'eat up' their greens in exchange for a 'naughty treat' later can warp their sense of taste for life, since it teaches them to 'treat pleasure and health as enemies'.[14] The early years are crucial too: fussy three-year-olds are likely to remain so for the rest of their days. Some aspects of taste do, however, alter with age: our sensitivity to bitterness decreases, for example, which is why the battles with sprouts tend to fade as we get older.

We depend on our sense of taste more than we realise. Those who lose it permanently report a profound sense of disorientation – a feeling of not quite being themselves. In a 2013 BBC documentary, the American chef Molly Birnbaum described her profound sense of distress at losing her sense of taste after a car accident. Tellingly, when her senses started to come back, she found they were powerfully linked to her emotions: the first flavours she could distinguish were those of rosemary, chocolate and wine, all related to happy memories from her youth.[15]

Intensely personal and emotional, our aromatic memories can be locked away for decades, which is why their sudden revelation can come as such a shock. But how do we store enough flavours to know what, for example, our classroom smelled like thirty years ago? The answer lies in numbers: we each possess around 40 million olfactory cells, 50 times fewer than a dog, yet still enough to distinguish between some trillion different smells.[16] This filing cabinet of odours, with its emotionally explosive payload, makes the perception of flavour one of the body's most complex and least understood functions.

Today, flavour perception is a rapidly expanding multi-disciplinary field in which chefs, psychologists and neuroscientists collaborate in order to unlock its secrets. For Professor Charles Spence of the Crossmodal Research Lab in Oxford, it is one of the most exciting new territories in biological science. 'Everything can alter our perception of flavour,' says Spence. 'This is a new science, and the possibilities are more or less infinite.'[17] In collaboration with chef Ferran Adrià (the celebrated Spanish 'father of molecular gastronomy') and the UK's similarly experimental Heston Blumenthal, Spence has explored how the shape, colour and texture of food, as well as the context in which it is eaten, can alter our perception of taste. One study undertaken with Adrià revealed that strawberry mousse served on a white plate tasted ten per cent sweeter than when served on a black one.[18] In another experiment with Blumenthal, Spence found that people eating the chef's fabled bacon and egg ice cream rated it more 'bacony' or 'eggy', depending on whether they were played soundtracks of pigs or chickens while eating. Now that the food industry is waking up to the possibilities, we can no doubt expect our sense of taste to be manipulated sometime soon, reaching parts of our psyches that other stimuli can't reach.

Taste is fundamental to our wellbeing, yet is in many ways our forgotten sense. So why do we ignore such a fundamental faculty? The answer is to do with our evolution. When our ancestors lived close to the ground, their sense of smell shaped their world, as it still does for dogs. When we started walking upright, however, our need to scan the horizon meant that sight became more important to us. Our awareness of smell receded, yet this primal sense remains deeply embedded, affecting us in ways of which we're largely unaware. Since the cerebral cortex (the human part of our brain) evolved at the junction between our ancient, reptile brain and the olfactory nerve, Harold McGee notes, it is fair to say that 'smell gave birth to the mind'.[19]

We Are What They Ate

> Animals feed: man eats.
> Jean Anthelme Brillat-Savarin[20]

What made us human? When did we start to think? Although answers to such questions are likely to remain elusive, what we do know for certain is that our ancestors ate before they thought. Three and a half million years ago our forebears were australopithecines: apelike creatures not unlike modern chimpanzees who were bipedal, social and good at climbing trees. They probably lived much as modern chimps do, moving in small groups, picking fruit and occasionally hunting, all with total unconcern for the meaning of life. Sometime around then, however, a spark must have lit, because scattered on the Ethiopian plains today are 3.4-million-year-old animal bones with clear cut marks, alongside deliberately sharpened flints. Some of our ancestors were making tools to butcher meat – technology was born.

The nature of those early tools is significant. Sharpened flints – and later spearheads and knives – tipped the balance of power between our diminutive forebears and their larger, faster prey. Like modern chimps, australopithecines were probably opportunistic hunters, catching small animals like monkeys that could be easily overpowered. To chase and dismember larger animals would have been beyond them. Armed with weapons and knives, however, a world of carnivorous opportunity opened up, and with it a rich new source of nourishment for their brains.

By 2.3 million years ago, australopithecines had morphed into habilines, whose brains ranged from 450 to 612 cubic centimetres (ours are around 1,400cc). Habilines still slept in trees, but ate a far meatier diet than their predecessors, probably pounding the flesh to make an early form of *steak tartare*. Half a million years later came *Homo erectus*, the first of our ancestors to walk upright and look recognisably human. We don't know whether they talked as well as walked, but with brains of 870 cubic centimetres, they had enough headspace to make arguably the most momentous breakthrough in human evolution, hailed by Darwin as 'probably the greatest ever made by man, excepting language'. Sometime between 1.8 million and 800,000 years ago, our ancestors worked out how to control fire.[21]

This changed everything. Mastery of fire allowed our ancestors to clear forests and attract grazing animals, to provide themselves with light and warmth, to ward off predators and sleep safely on the ground. It also created a focus (from Latin *focus* – fire) around which they could gather and socialise. Crucially, it also allowed people to start cooking their food, a development which, as Richard Wrangham argues in *Catching Fire*, made the biggest difference of all, since it altered the relationship between our two greediest organs, the guts and brain.

As anyone who has dozed off after Sunday lunch can testify, digestion is an energy-sapping process. Indeed, it accounts for 10 per cent of the body's resting metabolic rate (RMR), the energy needed to keep our bodies ticking over at rest. Meanwhile our brains, despite taking up just over 2 per cent of our body weight, consume one fifth of our

RMR.[22] Anyone tempted to replace visits to the gym with regular cryptic crosswords will, however, be disappointed: although intense mental effort can be exhausting, it only increases our energy use by a tiny fraction, since, whether or not we are consciously thinking, our brains must always be fully powered just to keep us alive. Inside our heads, despite appearances, the lights are always on.

With energy demands amounting to one third of the body's total output between them, our brain and guts are effectively in competition. If we want bigger brains, it follows that we need smaller guts: a piece of evolutionary logic known as the expensive tissue hypothesis.[23] It is here, Wrangham explains, that cooking made such a difference to our ancestors. Since cooked food is far easier to digest than raw, it allowed *Homo erectus* to divert energy away from digesting towards thinking. Instead of spending six hours a day chewing as modern chimps do, they could spend more time hunting and socialising. As their stomachs shrank and their brains expanded, they also became more adventurous in their diet, adding foods such as fish, packed with omega-3 fats, the brain's favourite five-star fuel. Cooking began a virtuous evolutionary circle that, around 200,000 years ago, produced the first of our own species, *Homo sapiens*.

Playing with Fire

> If all that changes slowly may be explained by life, all that changes quickly is explained by fire.
>
> Gaston Bachelard[24]

The significance of fire to humanity was not lost on the ancient Greeks, who told of how Prometheus had stolen it from the gods, a transgression for which he was eternally punished by Zeus. In this sense, the Greeks viewed fire much as Genesis treats knowledge in the story of Adam and Eve, a connection which the name Prometheus (Greek – forethought) seems to confirm. Fire and knowledge, for the ancients, were divine attributes that men had acquired, yet which they were unfit to possess. The ancients had a point: fire and knowledge are indeed assets that we find hard to live with. Knowledge granted us creativity, while fire has enabled us to put our ideas into practice. Together, they have made our lives infinitely richer and more comfortable, yet they have also led us to the brink of our own destruction. We struggle, it seems, to balance our own creative and destructive powers. Why might this be?

One clue comes from the famous image dubbed the *March of Progress*, first published in the 1965 Time-Life book *Early Man*. It shows human evolution through a series of ambling figures from left to right, from the 22-million-year-old, gibbon-like *Pliopithecus* through a series of less apelike, more upright creatures, to a spear-bearing early man and lastly a confidently striding *Homo sapiens*. Despite criticism that it portrayed the ascent of man as a triumphal procession, something its creator Rudolph Zallinger strenuously denied, the *March of Progress* is nevertheless fused in our imagination with the very idea of evolution.[25] This has proved a boon for cartoonists keen to draw the next figure in line: popular options include a man bent over a computer and a fat slob munching a hamburger. Humans, the cartoonists suggest, have over-evolved.

There is some truth in this. Obesity has overtaken smoking as the biggest killer in the West, and the US surgeon general recently warned that bad food habits and sedentary lifestyles could mean that the next generation of Americans would be the first to die younger than their parents.[26] Our post-industrial lifestyles aren't as good for us as we'd hoped, which, given our obsession with health and longevity and our unprecedented medical knowledge and access to drugs, is disappointing to say the least.

Part of the problem lies in what the *March of Progress* doesn't show: the changing environment that its figures inhabit. Of the 22 million years since *Pliopithecus* roamed the earth, 20 million rolled by before any humans showed up. Our ancestors tamed fire less than two million years ago and developed language within the last 100,000 years. We started farming just 12,000 years ago and building cities half as long ago again. We've had steam power for just 300 years, personal computers for less than fifty and the Internet for half as long as that. If one were to chart the great apes' technological prowess over time, the result would be a hockey-stick curve: a flat line millions of years long, followed by a gradual incline from the taming of fire, accelerating through the Neolithic period to reach an almost vertical trajectory today.

Our problem, in other words, is that we've used technology to turn evolutionary logic on its head. In the process of becoming human, we've stopped adapting ourselves to suit our environment, and have instead adapted our environment to suit us. This 'exo-evolutionary' approach worked fine for a while, yet its recent rapid acceleration has left our bodies out of synch with the world. If Darwin were alive today, he might say that we've made a fundamental evolutionary error. As he noted, it is the degree to which a species is suited to its environment, not its cleverness, that ensures its survival. In Darwinian terms, we've invented our way out of time. As the naturalist Edward O. Wilson put it, we've got 'Stone Age emotions, medieval institutions, and god-like technology'.[27] No wonder we struggle to live in the modern world: we come from another planet.

Alpha and Omega

All flesh is grass.
Isaiah[28]

Nowhere is this temporal mismatch more evident than in the way we eat. During the past two centuries, industrial farming has lifted the scourge of hunger from great swathes of the developed world. Yet this achievement is not quite all it seems. The industrial acceleration of food production has come, as we have seen, at near-incalculable ecological and human cost, with the added catch-22 that the population explosion it has allowed has vastly increased our global demand for food – a sort of reverse Malthus, if you will. Then there is the quality of the food itself. One might expect denizens of the world's leading industrial nations to be the best nourished on earth, yet, in many respects, we are the worst.

What does it mean to eat well? For the temperate-minded Greeks, it was a case of *meden agan* (nothing in excess), and certainly balance and moderation are two keys to a healthy diet. Even without knowing what's in our food, such an approach is likely to deliver a reasonable balance of the three macronutrients that our bodies need: fats, proteins and carbohydrates. Obtained mostly from plants, carbohydrates provide energy and fibre to help us digest, while fats and proteins come from animals and plants to provide materials our bodies need to build and repair themselves. Since fats and proteins also provide energy, it's possible to live with very few carbs in one's diet – as the Inuit have done with great success, and Atkins dieters less so. None of us, however, can live without protein and fat.

Food's passage through the body is one of progressive metamorphosis, as our digestive system breaks it down into its constituent parts and reassembles them into forms that our bodies can use. Carbs are mined for glucose, the body's preferred energy fuel; proteins become amino acids, used to make and repair cells; fats break down into fatty acids, vital to the structure of the brain, liver and nervous system. Some nutrients we need can be synthesised by the body, while others – called 'essential' for this reason – can't. Eight of the twenty amino acids our bodies need are essential, as are two fatty acids: omega-3 alpha-linolenic acid and omega-6 linoleic acid. Although no specific carbohydrates are essential to the body, shortage of them can overburden the liver, which must then work overtime to manufacture glucose from fats and proteins, while lack of fibre can cause digestive problems. Lastly, excess energy from any of these processes can be stored by the body, as few of us need be reminded, in the form of adipose tissue, aka fat.

In addition to the big three macronutrients, our bodies need around forty essential minerals. Of these, seven (calcium, magnesium, phosphorus, potassium, sodium, sulphur and chloride) are needed in relatively large amounts, while the remaining 'trace' elements (including iron, cobalt, copper, chromium, iodine, manganese and zinc) are needed in tiny quantities of a few grams or less. Although present in minuscule amounts, these trace elements are as vital to bodily function as the rest. The same goes for vitamins, a group of organic compounds needed in even tinier doses – sometimes just a few hundredths of a gram – the absence of which can lead to sickness or death, as ancient mariners stuck on long sea voyages discovered to their cost.[29]

Mariners' suffering highlighted the importance of a balanced diet long before any such concept existed. Not all healthy human diets are balanced, however, as traditional peoples like the Inuit and Maasai prove: the former flourish on a diet of seal, walrus, fish, birds and eggs plus a few tubers, roots and berries, while the latter live mostly on blood and milk from their cows. One of the luxuries of being human is to possess a remarkably adaptable digestive system: we have devoured some 80,000 different plant and animal species in our time, of which 3,000 have been in widespread use.[30] Our bodies have adjusted to living in some pretty inhospitable places too. In order to survive on their low-carb diet, for example, the Inuit have developed enlarged livers that help them convert protein and fat into energy. Northern Europeans also enjoy a dietary advantage absent in most other humans (but also present in the Maasai): a tolerance for lactose, the carbohydrate found in milk.[31] Due to a genetic mutation first thought to have occurred among a group of Polish or Turkish herders around 5,000 years ago, the milk-digesting enzyme lactase (which in most humans switches off after weaning) stays active into adulthood. The benefits of being able to drink milk meant that the mutation spread quickly, in a sort of dairy-led form of Darwinism.[32]

After the invention of cooking, the greatest transition in the human diet came with the advent of farming. As our ancestors swapped their regime of meat, fish, nuts and berries to one based mostly on grains and pulses, there was a seismic shift, not just in the content of human food, but in its diversity. Today just three plant species – wheat, rice and maize – are the staple of three quarters of the global population.[33] Apart from the lack of resilience that such an approach to feeding ourselves represents, there is the question of whether such a narrow diet can be good for us. While some argue that our bodies are not suited to eating grain and that we should revert to a so-called caveman, or Paleo, diet, the reality is that leaving aside the question of how we would subsist without cereals, our flexible guts mean there is no such thing as a 'natural' human diet, just countless ways of eating better or worse.[34]

The main problem we face when it comes to eating, as Michael Pollan points out in *The Omnivore's Dilemma*, is choice. Humans are nothing if not adaptable, which is why it's ironic that our food system has become so one-dimensional. We may be confronted with a choice of fifty breakfast cereals in a supermarket aisle, yet in the end, all are just cereal – modified, reduced or enhanced in various ways. As Pollan points out, the healthiness of our diet doesn't necessarily depend on the range of things we are offered, but in our food's inherent richness. The fact that you are what you eat means, he notes, that 'you are what what you eat eats too'.[35]

The Inuit thrive because the seal and walrus on which they dine eat fish raised on algae and phytoplankton bursting with omega-3 fats and vitamin C. Their diet may seem simple, yet it draws from the greatest reservoir of nourishment on our planet, the ocean. Knowing how to eat well in any location is what local food cultures have always done, and physiologies have adapted to suit. The key difference between the Inuit and the British seafarers who strayed so fatefully into their territory in 1845 was that the locals drew their sustenance from the depths of the sea, while the visitors merely floated upon it. For many, this was the difference between life and death.

Against the Grain

> Let food be thy medicine and medicine be thy food.
>
> Hippocrates

As post-industrial humans, we certainly *ought* to know how to eat. Traditional diets contain all the nutrients we need; if they didn't, after all, we wouldn't be here. As super-nerd Sheldon Cooper quipped in the US comedy *The Big Bang Theory*, taking multi-vitamins on top of the average healthy diet is merely buying 'ingredients for very expensive urine'.[36]

As long as we eat a varied and balanced diet, we should be fine. Yet that is perversely what our industrial food system makes it harder for us to do. Take carrots, for example. Even the most ardent advocate of industrial farming admits that a carrot pulled straight out of the ground and nibbled on the spot tastes better than one that has been bagged, gassed, bleached or frozen to withstand the thousand-mile merry-go-round of modern food logistics that brings it to our door. Eating processed and packaged food is part of the price we pay for living far from where it is produced. More worrying, however, is the question of what is in the carrot itself. Over the past fifty years this has radically altered, as farmers have turned increasingly to chemical fertilisers and pesticides, depleting once rich, fertile soils. Records from the British Medical Research Council from 1940 to 1991 reveal that during this period carrots lost 75 per cent of their copper and magnesium, 48 per cent of their calcium and 46 per cent of their iron.[37]

Similar things have happened to beef, the food once so beloved by the British that it earned us the French sobriquet *les Rosbifs*. Traditionally reared on pasture (of which the maritime UK has an ample supply), beef is a superfood full of minerals, vitamins and complex omega-3 fatty acids. As grass-munching ruminants, cows could be said to rival dogs for the title of man's best friend, transforming inedible (to us) cellulose into nutritious beef and milk. Cows can do this thanks to their rumen, essentially a large fermentation tank that can break down complex molecules to create high-grade foods that we can digest. Since most cattle these days are fed on grain rather than grass, however, this beautiful synergy is lost. Cattle are not suited to eating grain: it gives them a permanent case of indigestion and sends toxins into their bloodstream that antibiotics are used to quell. Their fast-food diet also means that instead of producing lots of omega-3-rich muscle, they instead produce omega-6 fats.[38]

This is bad news for both cows and us, since omega-3 fats are rapidly vanishing from our diet. Found mostly in green plants and fish oils, omega-3s are a superfood vital to brain function, vision and anti-inflammatory action. Although we also need omega-6s (which perform complementary roles in the body), the latter are overabundant in our industrial diet. Since both types of fat compete to be absorbed by the body, a surfeit of omega-6s increases our deficiency in omega-3s. Ideally, we would consume a ratio of 1:1 (as our forager forebears did), and a ratio of 4:1 is still considered tolerable, but in the West the ratio is often as high as 10 to 1, enough to threaten our mental and physical health. One recent survey in the US found that 60 per cent of the population was omega-3 deficient, and 20 per cent had levels so low they were undetectable.[39] According to Oxford University Professor of Physiology

John Stein, 'The lack of omega-3s in our diet is going to change the human brain in ways that are as serious as climate change.'[40]

Beef and carrots are just two examples of foods that industrial farming has denatured. Yet fresh ingredients like these that your granny might have cooked with are just the tip of the industrial food iceberg. A 2018 study of nineteen European countries by the journal *Public Health Nutrition* found that more than half of the food bought to eat at home in the UK consisted of 'ultra-processed' foods, made in factories using industrial ingredients unknown in a domestic kitchen.[41] Predictably, the UK came top of the chart, with 50.7 per cent of the average shopping basket consisting of ultra-processed foods, compared to just 14.2 in France and 13.4 per cent in Italy.[42] As we abandon cooking from scratch in favour of such industrial pseudo-foods, our wallets aren't the only things being hit. One 2018 study led by researchers based at the Sorbonne in Paris found a direct link between the eating of such ultra-processed foods and the incidence of certain types of cancer.[43]

The quality of our food has changed beyond recognition in the past fifty years, and some of it is doing us actual harm. So how, without a portable chemistry set or opting to grow our own, can we tell whether the food we eat is good for us or not? In an ideal world, we'd all cook from scratch using fresh ingredients and buy our food directly from trusted local producers and transparent supply chains, circumventing the industrial system altogether. That is what farmers' markets and organic box schemes effectively do, demand for which has already forced some British and American supermarkets to up their game. For most of us, however, time and cost remain top priorities when it comes to food.

As the Slow Food founder Carlo Petrini has pointed out, good food doesn't have to be expensive: Italian peasant cuisine – so-called *cucina povera* – boasts some of the finest dishes in the world.[44] Eating in such a way does, however, require knowledge, time and skill, as well as access to trusted markets for the right ingredients – in other words, it requires a traditional food culture. Eating well in places where such traditions persist is easy, yet it is conversely hard to do so where no such culture exists; where, for example, people live in food deserts – poor urban neighbourhoods where no fresh food is available.[45] For those who live in a depleted food culture, the barriers to eating well don't just come down to time, money or skill – they include resistance from their own bodies. When Jamie Oliver took some disadvantaged schoolchildren to pick fresh strawberries, for example, many of the children gagged at the unfamiliar taste. Having never encountered fresh fruit before, the flavour revolted them.[46] In the UK, this phenomenon is far from new: in his 1937 book *The Road to Wigan Pier*, George Orwell noted how an unrelenting diet of processed foods had 'industrialised' the nation's sense of taste:

> The English palate, especially the working-class palate, now rejects good food almost automatically. The number of people who *prefer* tinned peas and tinned fish to real peas and real fish must be increasing every year, and plenty of people who could afford real milk in their tea would much sooner have tinned milk.[47]

As the world's first industrial nation, the UK has a long history of eating badly. Our Victorian enthusiasm for corned beef, condensed milk and tinned peaches gave birth to an industry predicated on selling cheap processed food to the masses. Today, with Pot Noodles, Pop-Tarts and Cheesy Wotsits occupying the middle aisles of our supermarkets, the habit remains ingrained. The fact that organic food exists as a rarefied subcategory says it all.

Fat Land

> Man shall not live by bread alone . . . he must have peanut butter.
> US President James A. Garfield

Few of us are well equipped to negotiate the minefield of industrial food. Most of us have no clue as to what our bodies need in order to be healthy; that is what traditional food cultures used to tell us. The roasts, stews, chops and fish and chips of my 1960s childhood may not have been exotic, but at least my parents cooked the meals from scratch and we knew what we were eating. Now that supermarkets have replaced Mum and Dad as primary food providers, all bets are off. Supermarkets don't give a stuff about whether we eat healthily; their goal is to make as much profit from us as possible. As far as they are concerned, we can have as many pizzas and packets of crisps as we like – which, given the chance, many of us will do.

This is not our fault. Our bodies are programmed to feast when times are good – they don't know we're not still out hunting and gathering. As Marion Nestle points out in her 2002 book *Food Politics*, the fact that the US food industry produces twice as much food as Americans can safely eat poses a problem for producers as well as consumers, since it requires companies to pressure-sell their food. 'Supersize' promotions and BOGOFs (buy one get one free) are the result, along with billions spent on lobbying Congress, sponsoring 'friendly' research and marketing directly to kids.[48] Since junk foods yield by far the highest profits, that is where the bulk of the budget goes. In 2012, the US fast-food industry spent $4.6 billion on such advertising; by comparison, the Department of Agriculture (USDA) spent just $6.5 million promoting fruit and vegetables.[49]

No wonder that 70 per cent of Americans are overweight and 40 per cent obese.[50] No surprise, either, that a disproportionate number of these come from disadvantaged groups. The poorest in society have rarely eaten well, yet in an ironic modern twist, many of the 47 million US citizens who depend on government food stamps live in food deserts and are thus forced to spend them on junk food. As a recent Gallup poll suggests, many Americans feel trapped in their food choices: although 76 per cent of respondents said they thought fast food was 'not too good' or 'not good at all for you', 47 per cent said they ate it at least once a week.[51]

New World Syndrome

Like an unstoppable tide, US food culture is spreading across the world, creating havoc wherever it goes. Among the first to feel its effects were the Marshallese, whose Pacific archipelago was captured by US forces during the Second World War. Before the war the Marshallese were mostly hunter-gatherers, with a diet that reads like a nutritionist's dream: fish and shellfish, coconut, breadfruit, leafy greens and *pandanas*, fibrous fruits high in carotenoids (powerful antioxidants). When the US started using Bikini Atoll as a nuclear testing site, however, most of the population were moved to the capital, Majuro, where they were forced to live on imported US foods. With a diet that now consisted of white rice, corned beef, tinned vegetables and sugary drinks, the Marshallese were soon among the most badly nourished people on the planet. Today almost 75 per cent of women and 50 per cent of men are overweight or obese, and nearly 50 per cent of adults over the age of thirty-five suffer from diabetes, with related amputations accounting for one half of all surgeries performed on the islands. So dire have the effects of their diet been that they have earned their own label: New World Syndrome.[52]

Where the Marshallese had no choice but to go, others have chosen to follow. Another fast-food invasion occurred in the Middle East on the back of the 1991 Gulf War. Chains including Burger King, Pizza Hut and Taco Bell that came to feed US troops were met with such enthusiasm by locals that they stayed on for the peace, producing hybrids such as Pizza Hut's bestselling Cheeseburger Pizza Crust to satisfy local demand. Today, 88 per cent of Kuwaitis are overweight or obese, making Kuwait the first nation to overtake the USA as the fattest on earth.[53]

Despite such portly portents, fast food continues to sweep all before it, even in such renowned culinary nations as France, Italy, India and China. To begin with, there was some resistance: the 1986 arrival of McDonald's in Rome so incensed Carlo Petrini that he set up a stall opposite handing out 'Italian slow food' – home-made pasta – to passers-by; three years later, he founded the Slow Food movement. Similarly, when the golden arches sprouted in the Pyrenean town of Millau in 1999, sheep farmer Jose Bové threw bricks through its windows in protest at the 'gastronomic imperialism'.[54] Such resistance has proved futile, however: McDonald's remains the largest burger chain in the world, operating 36,000 restaurants in 119 countries, including 1,400 in the birthplace of gastronomy itself.[55] The French are now the second-biggest consumers of Big Macs outside the United States. Such is the enthusiasm for 'McDo', indeed, that in 2014 residents of Saint-Pol-sur-Ternoise marched to demand the opening of a branch of McDonald's in their town.[56]

Why would anyone want to swap *boeuf bourguignon* or *lasagne al forno* for a Big Mac? Lashings of salt, sugar and fat are clearly part of the appeal, as is the cultural cachet. As Andy Warhol pointed out, part of fast food's appeal lies in the way that its commercialised, one-size-fits-all culture slots seamlessly into the American Dream:

> What's great about this country is that America started the tradition where the richest consumers buy essentially the same things as the poorest. You can be watching TV and see Coca-Cola, and you know that the president drinks Coke, Liz Taylor drinks Coke, and just think, you can drink Coke, too. A Coke is a Coke and no amount of money can get you a better Coke than the one the bum on the corner is drinking. All the Cokes are the same and all the Cokes are good. Liz Taylor knows it, the president knows it, the bum knows it, and you know it.[57]

Few have exploited this subliminal message more skilfully than billionaire President Donald Trump, whose habit of ordering in hamburgers to the White House endeared him to his fans as a 'man of the people' in a way that few other acts could have done. For a new generation coming of age in nations such as India and China, fast food catches the upbeat spirit of America's original teenage paradise: the *Happy Days* of the 1950s replayed for the digital age. Indian street-food sellers have even restyled their operations to resemble US fast-food joints, Americanising their dishes to suit. If you want to eat *Bhel puri* and *Aloo tikki* the traditional way, you had better get yourself off to India quick.

By offering us a way of eating devoid of rules and tradition, fast food seems to offer a form of freedom. Food without limits or responsibilities, it echoes the medieval dream of the Land of Cockaigne, an edible world dripping with things to be eaten, itself a rehash of the Garden of Eden. Thanks to industrial food, many of us now live in such a land, one in which we can graze all day if we want to, without worrying too much about where our food came from or what the real landscape looks like behind the facade.

Pleasure Dome

> I can resist everything except temptation.
> Oscar Wilde[58]

Epicurus observed that humans are hedonists and that satisfying bodily needs was one of the ways we got our regular fix. He also noted that when we overindulge, our pleasure soon turns to pain. Now neuroscientists are starting to understand why. In the 1950s, American scientists James Olds and Peter Milner discovered separate areas in the brain associated with pleasure and pain, calling them the reward and punishment centres. Today we know that these centres are linked to other parts of the brain via neural pathways down which neurotransmitters such as dopamine travel. When we are stimulated by certain foods or drugs, our dopamine levels shoot up, causing our reward systems to activate, a process that, thanks to modern brain scanners, we can now witness in real time.

Crucially, our reactions to such stimuli are involuntary. Although we experience taste in our conscious brain – the prefrontal cortex – our responses to food are driven by far older, subconscious parts: the reptilian brain at the top of the spinal cord and our limbic system just above that. Together, these regions are responsible for much of our motivation, prompting us to seek food, safety and sex (not necessarily in that order). When we're under threat, our emotional centre (amygdala) triggers our hormone centre (hypothalamus) to release cortisol into the bloodstream and put us on high alert. When we are pleasantly stimulated, on the other hand, our pleasure centre (nucleus accumbens) releases dopamine into the brain's motivational centre (striatum), prompting us to seek more of whatever it was we just had. Meanwhile, the memory bank (hippocampus) faithfully records our reactions and stores them away, which is why, fifty years on, we might suddenly go misty-eyed over a madeleine.

While our reptile brain and limbic system are busy pumping out hormones to swing our mood this way or that, our prefrontal cortex remains magisterially aloof. Since our sense of self resides in the prefrontal cortex, this set-up gives us something akin to a split personality when it comes to pleasure and pain. Of all the parts of our brain that light up when we eat a chocolate doughnut, only the prefrontal cortex asks itself whether a second one might be a good idea. It is, in short, the only part of us that wrestles with temptation.

Gut Feeling

Fortunately for us, our brain isn't the only part of our anatomy that regulates how we eat; the gut also plays a major part. Our gastrointestinal tract might have shrunk at the expense of our brain, but it's still a formidable organ: nine metres long, with a surface area of 4,500 square metres (equivalent to seventeen doubles tennis courts), it was until recently one of the least understood – and least valued – parts of our body. Not any more: recent advances in microscopy have revealed the gut to possess 100 million neurons and thirty neurotransmitters, effectively making it a second brain as big as a cat's that works in tandem with the one up top.[59] This so-called gut–brain axis forms the core, not just of how we eat, but of how we sense, and make sense of, the world.

Like our limbic system, our gut plays with our motivational circuits, releasing hormones that promote or inhibit our brain's receptiveness to pleasure and pain. When the gut starts to empty, it releases 'hungry' hormones including ghrelin and PYY to whet our appetite and prompt us to search for food; once we have eaten, it sends leptin and serotonin to reverse the process, reducing our dopamine receptiveness and thus our pleasure in eating. The result is a hedonic cycle that takes us from pain to pleasure and back again in an eternal pattern synchronised with our diurnal rhythm. Our second brain thus commands our first, not just in the matter of when to eat, but in a rhythmic sequence that profoundly affects our mood. All of which explains why listening to your gut is more than mere metaphor – and why, if you want a pay rise, it's better to ask the boss after lunch.

For most of history the system worked a treat. Hunger was a default condition for our forager forebears, putting their bodies in a more or less constant state of arousal, driving them to seek food. Eating would have given them intense pleasure, and since overeating was rare, that pleasure would have lasted for the entire meal. Foods high in sugar such as honey were hard to get and highly prized: contemporary hunter-gatherers will risk nasty bee stings, even their lives, climbing high into trees to obtain the precious substance.

Today, with similar bodies yet radically altered lifestyles, our reward systems struggle to cope. Rather than climb trees for honey these days, we simply lob a jar into our trolley – an arrangement that, for all its advantages, robs us of much of the satisfaction of eating. In our over-stuffed world, enjoyment falls off pretty rapidly, which is probably why nibbling cuisines such as tapas, dim sum and sushi have become so popular: by providing tiny portions and lots of variety, they prolong our pleasure for longer than the pies and stews that once filled our labour-ers' bellies.

Constant over-eating can also have more serious consequences, over-riding the brain's responsiveness to signals from the gut, leading to a downward spiral similar to addiction.[60] In an experiment with rats, the neuropsychologist Paul J. Kenny found that those given access to a 'rodent cafeteria' stocked with sausages, chocolate and cheesecake would carry on eating, despite being administered periodic electric shocks. Rats fed on a normal diet, meanwhile, scrambled away to safety. The pleasure of eating cheesecake, it seems, was worth the pain: Kenny's 'cafeteria' rats became so besotted with their food that they would literally eat themselves to death. As he put it, 'Their hedonic desire overruled their basic sense of self-preservation.'[61] Indeed, says Kenny, his rats behaved in exactly the same way as those in a parallel experiment – and they were hooked on cocaine.

We're not rodents, yet it seems that a rational mind is little defence against the addictive lure of foods like cheesecake. Why should this be? According to Kenny, it's because such foods contain combinations that would never occur in the wild. In nature, one finds plenty of sugar and fat, yet never the two together. It is only we humans – cooking animals – who have worked out how to make fatally delicious food (think salted-caramel ice cream). Eating such treats on a regular basis can hijack our brain's reward system, flooding it with so much dopamine that our pleasure responses start to wane, meaning we must eat ever more to get the same hit as before. This, according to Kenny, can trap those who regularly overeat (i.e. the obese) in a condition very much like tolerance: 'As with alcoholics and drug addicts, the more they eat, the more they want.'[62]

The French Paradox

Millions of us live in an obesogenic world, so why aren't we all fat? Part of the answer is that our ability to live well in the Land of Cockaigne differs as much as our capacity to be happy. Some of us are simply more interested in food than others. While some (I confess) tend towards the Labrador end of the spectrum, others can be so uninterested in food as to forget to eat. The difference is partly down to genetics. One long-term study on twins run by the genetic epidemiologist Professor Tim Spector at King's College London found that children's genes had a 25 per cent effect on their likelihood of being obese in adulthood.[63] There is also the matter of how active we are: Olympic swimmer Michael Phelps reportedly scoffed 12,000 calories a day; once he got out of the pool, however, he soon piled on the pounds.

A plethora of factors – body type, personality, habit, circumstance, education and genes – play a part in determining how fat we're likely to be. For the past several decades, social disadvantage has been a major factor in the likelihood of becoming overweight or obese. Recently, however, social status appears to be losing its significance. Whatever it is that is making us fat, it is now doing so across the social spectrum.[64] In Britain, only two social groups (rich women and poor men) are bucking the trend. This finding puts paid to the idea that fat people simply lack the willpower to diet. Although that might be true in certain cases, many are simply victims of living in an obesogenic world.

What makes some societies more obesogenic than others? Why, for instance, does living in Britain make you fat, while living in France – at least for now – does not? The answer comes down to the ways in which industrialisation dismantles traditional food cultures. One of the reasons why Britons are fat is that, due to industrialisation, we abandoned our local food customs long before anyone else. French food culture, meanwhile, although clearly under threat from McDo, is still relatively intact, which affects not just the way the French eat, but how they live.

Famously, most French take their food very seriously: the Michelin Guide, for example, began because people insisted on eating well even when on the road. Quality, provenance and seasonality remain paramount in France, and high-quality independent food shops are still common. There is a 'right' and 'wrong' way to cook most dishes, which must be served *comme il faut*. Unlike the British and Americans, the French also take plenty of time to enjoy their food: stroll through Paris at midday during the week, and you'll find restaurants packed with workers enjoying a good lunch, rather than the snatched sandwich more common in the US and UK.[65] Despite spending more time on food and eating than any other Western nation, however, the French aren't fat: they have one of the lowest obesity rates in Europe. There is also the so-called French Paradox: the fact that, while famously relishing cheese and cream, the French have an enviably low rate of heart disease.

As anyone who has eaten in France will know, the paradox is no such thing. When a groaning cheeseboard rolls up at a restaurant, most French will take just a tiny amount of two or three varieties. Lingered over with a glass of wine, however, the speck of often eye-wateringly pungent cheese goes a long way. As with all traditional food cultures, the rules determine not just what one eats but how.

As Paul Rozin, Abigail K. Remick and Claude Fischler showed in a 2011 study of French and American food cultures, such attitudes spread far beyond the table.[66] The French, for example, enjoyed their meals without any sense of guilt, whereas for Americans such enjoyment was seen as a guilty pleasure. Such a difference, the team suggested, may be due to the historical role of Catholicism and Protestantism in the two cultures:

> The Protestant tradition is characterized by a greater emphasis on self-discipline, on control of the body, and on individuality. Pleasure

is more likely to be confounded with sin and guilt among Americans
... [who] believe it is the individual's responsibility to remain healthy,
fit, and slim, and if the individual fails then he or she can be deemed
irresponsible. Following from this, Americans have attached a sub-
stantial moral component to health, dieting, and fat.[67]

Eating was also a far more social activity in France, the study found,
reflected in a greater acceptance of eating in the 'correct' way at the
expense of individual choice:

France has a longer history and a much better defined cuisine and
sense of the role of food in life than does America. One consequence,
we maintain, is that the French seek less micro-variation in their cuisine, since there is more likely to be an accepted (best) form for any
dish or food. While Americans expect to choose whether to have
French fries, mashed potatoes, baked potatoes, or home-fried potatoes with their steak, the French assume that steak goes with *frites*.[68]

The fact that traditional food habits remain relatively intact in France
is reflected in the language. There is no equivalent concept to comfort
food, for example, nor any term that equates to our all-purpose word
for food. As Fischler points out, no French speaker would talk of *ma
nourriture préférée* (my favourite food), but would rather refer to something specific: *mon plat préféré* (my favourite dish), *ma cuisine préférée, ma
pâtisserie préférée* and so on.[69]

Such differences in food culture, the team found, were mirrored in
other areas of life. When shopping for clothes, for example, the French
were happy to follow the advice of experts, while Americans wanted to
choose for themselves. With respect to both food and clothes, the
French sought quality over quantity, while Americans preferred the
reverse: a contrast reflected in the relative expense and smaller portion
sizes of French meals compared to American ones. The authors also
found a sharp contrast in attitudes towards comfort: while travelling, for
example, Americans valued comfortable rooms with good beds and air
conditioning, while the French were more concerned with local pleasures such as going to the theatre than with hotel amenities.

Comfort and Joy

Less is more.
Mies van der Rohe[70]

What can we learn from these two contrasting cultures? French values are Old World and traditional, while American ones are New World and consumerist. French culture is more social and contextual, while Americans are more individualist and moralistic. The French seek quality, expertise and pleasure, while Americans prefer quantity, choice and comfort. So which set of values is more suited to a happy life? The question matters, because it is the latter set, in the form of consumerist capitalism, that is going global.

As the Hungarian-American economist Tibor Scitovsky pointed out in his 1976 book *The Joyless Economy*, one key difference between traditional and consumerist societies is the way in which they manage pleasure. While traditional cultures contextualise and delimit pleasure, consumerist ones aim to deliver it 24/7. Although the latter approach may appeal more to our reptilian brains, said Scitovsky, it can also reduce our sense of enjoyment, by messing with our body's motivational systems: 'We feel ill or well, we experience pain or pleasure, and when we do, we are sensing our level of arousal. Moreover, because we seek pleasure and try to avoid pain, the concept of arousal is central to the explanation of behaviour.'[71]

As long as it doesn't last too long, said Scitovsky, such a state of arousal can actually be enjoyable. Our senses are heightened, so that when our desires are finally met, we experience a surge of joy that slowly subsides into comfort and contentment. For the time being, we are satisfied – until the whole cycle starts again.

Our ancestors would have been familiar with such a rhythm. In Aristotelian terms, contentment (comfort) is the goal of the human hedonic cycle: the mean between desire and excess that our reward systems are calibrated to deliver. But here's the rub: if joy is what we seek rather than comfort, we must allow our arousal levels to build, since our greatest pleasure occurs just as they border on the unpleasant. First noted by the German experimental psychologist Wilhelm Wundt in 1874 (and thus known as the Wundt curve), this phenomenon graphically illustrates our modern dilemma. In order to enjoy life to the full, we need our wants to be postponed, rather than instantly gratified: to experience pleasure at its peak, we must work towards it and look forward to it. However, as Scitovsky notes, such postponement is the opposite of what consumerist culture is geared up to provide. We miss out on joy, because our needs are met too easily.

Comfort and joy, it turns out, are to some extent mutually exclusive. If we want to experience joy, we must be prepared to sacrifice some comfort, yet our very idea of a good life – embodied in the notion of progress – is to ratchet up the latter. In a consumerist society, said Scitovsky, we are constantly forced to choose between comfort and joy, yet this isn't clear to us, since our gains in the former (such as when we snack) are often present and immediate, while the consequent loss (no appetite at dinner) only becomes apparent later.[72] Instinctively, we accept ever-increasing levels of comfort, without realising that we are pushing joy ever further away.

Fruit and Nuts

> The decline of a nation commences when gourmandising begins.
> John Harvey Kellogg[73]

One result of modern food culture, in which infinite choice is offered and consumers are assumed to know best, is a deep confusion about what and how to eat. Such perplexity has led inexorably towards faddism, as people search for the kind of knowledge and guidance that their families might once have given them as part of their heritage.

As the world's greatest cultural melting pot, nineteenth-century America was the natural crucible for such faddism, as quacks and cranks of all kinds found willing disciples among the millions of migrants eager to learn how to eat in their newly adopted country. The gurus' job was made easier by the fact that very little was then known about nutrition. For such an important science, nutrition had a surprisingly late start: the key macronutrients of fats, carbohydrates and proteins were only identified by the English chemist William Prout in 1825, and the first to postulate how they worked was none other than Justus von Liebig, who in 1842 diverted his attention away from fertilisers for long enough to suggest that proteins were used by the body to form tissues and muscle, while fats and carbohydrates were burned for energy. Although wrong in certain respects, Liebig's was nevertheless the first attempt at a concerted theory of nutrition.

In the US, however, the quest for an ideal diet was becoming fused with another characteristic American pursuit: salvation. Presbyterian minister Sylvester Graham set the trend, advocating whole grains as the foundation of a moral and healthy diet on the basis that since milling grain 'put asunder what God has joined together', it was sinful.[74] In 1829 Graham launched the world's first individually formulated vegetarian diet, consisting of whole grains, fruit and vegetables, supplemented by small amounts of fresh milk, cheese and eggs. Most modern nutritionists would approve of Graham's diet, yet the minister's motives were moral, not medical: he objected to eating meat on the basis that it led to 'impure thoughts'. Despite being ridiculed for this obsession, Graham nevertheless became a cult figure. In 1863, a fan by the name of James C. Jackson had the idea of grinding up his Graham bread into tiny chunks and re-baking them to make Granula, a combative concoction meant to be eaten softened in milk. Rebranded Grape Nuts, they were the world's first breakfast cereal.

Despite Graham's fame, it is another name, Kellogg, that is most likely to appear on our breakfast tables today. Another deeply religious man (a Seventh Day Adventist in his case), John Harvey Kellogg believed, like Graham, that the prevailing American diet was damaging to both health and morals, and that the greatest evil corrupting humanity was lust, as he outlined in a thundering tract named *Plain Facts for Old and Young*: 'The science of physiology teaches that our very thoughts are born of what we eat. A man that lives on pork, fine-flour bread, rich pies and cakes and condiments, drinks tea and coffee and uses tobacco, might as well try to fly as to be chaste in thought.'[75]

In the matter of diet, Kellogg was even stricter than Graham: pretty much anything other than whole grains, nuts, fruit and vegetables was off, with the sole exception of 'a Bulgarian milk preparation, known as yoghurt'.[76] Kellogg believed that his diet came directly from God, a doctrine that he preached in his role as director of the Adventists' Battle Creek Sanitarium. Patients at the 'San' were given strict, individualised diets and were put through breathing exercises and mealtime marches to promote digestion, as well as being encouraged to chew food until it slipped down their throats of its own accord. None escaped Kellogg's favourite treatment: a daily ritual to purify the gut of 'harmful bacteria' consisting of the administration of the 'Bulgarian milk preparation' at both ends.

Despite his lack of scientific credentials, Kellogg became a leading nutritional expert of his day, counting John D. Rockefeller and Theodore Roosevelt among his clients. But his name would probably not resonate today had he and his more business-like brother Will not come up with the idea, in 1895, of rolling cooked wheat kernels and baking them until they went brown and crispy, eventually switching to corn and adding sugar to create the world's most famous breakfast cereal, cornflakes.[77]

The New Nutrition

> The emperors of nutrition have no clothes.
>
> Michael Pollan[78]

With their cultish followings, lack of scientific knowledge and moralistic labelling of foods as good and bad, Kellogg and Graham effectively founded modern food faddism. Although today's food gurus are more likely to be Instagram-friendly young women than craggy priests, the phenomenon remains the same. As traditional food cultures have continued to fragment, opportunities have opened up for anyone with a dietary message to sell it to a credulous public.

The next stage in the evolution of what Michael Pollan has dubbed the Age of Nutritionism was the food industry's realisation of the phenomenal potential of nutritional science to sell its wares.[79] As home economists, biochemists and journalists turned their attention towards this oddly neglected field during the 1920s and 30s, Americans were subjected to a welter of new information, learning that their meat-and-two-veg dinners were in fact proteins, carbs and fats and being told that, rather than enjoying their apple pie and cream, they should be counting the calories.

Diets were no longer aimed at curbing lust and serving God, but rather at health, longevity and the very thing that Graham and Kellogg had been so keen to stamp out: sex. Each new nutritional discovery was hailed as a miracle cure and hastily incorporated into the latest processed foods. Vitamins, first isolated by the Japanese scientist Umetaro Suzuki in 1910, were received with barely controlled hysteria, with extravagant health claims made on their behalf such as curing stomach ulcers and tooth decay or boosting energy and brainpower. With 1920s flappers signalling a new slimline ideal of female beauty, millions of American women began fretting about their size, an obsession that an erroneous table of 'ideal' weights published by Metropolitan Life in 1942 (based on data from twenty-five-year-olds) did little to dispel, branding as it did half the nation overweight.[80] When Dior's wasp-waisted New Look arrived in 1947, the US diet industry was set to explode.

Metrecal, a 225-calorie diet drink launched in 1959, was the first food product to exploit the yawning gap between aspiration and reality. Flavoured with vanilla and consisting of skimmed milk, soybean flour and corn oil fortified with vitamins and minerals, it was marketed as part of a new sexy lifestyle that, as *Forbes* magazine noted approvingly, moved it 'out of the medicine cabinet toward the kitchen, the patio, the pool'. By 1960, Metrecal was a mania, with Greek and Saudi royals among its fans, as *Time* magazine recorded: 'Across the nation last week, drugstores and supermarkets were clamoring for fresh carload deliveries to accommodate the growing hordes of Schmoo-shaped addicts who were insisting on guzzling their way to the vanishing point.'[81]

Metrecal was soon overtaken by a bevy of rivals, yet it had proved that there was a mint to be made out of selling people shots of gloop on the premise that it would make them slim and successful. Around the same time, the US diet industry received a further boost in the form of an American pathologist by the name of Ancel Keys. During the 1950s, Keys had conducted an international study that he claimed showed that saturated fats led to raised cholesterol and heart disease.

Despite flawed evidence (Key's subjects were all male and mostly smokers), Keys persuaded the American Heart Association to accept his ideas, and in 1961 the AHA recommended that Americans cut back on saturated fats.[82]

Rarely has official dietary advice had a graver effect upon a nation's health. As butter gave way to low-fat spread and bacon and eggs were replaced by orange juice and cereal, several terrible things happened to the American diet. First, it got a lot less tasty; second, its carbohydrate content, much of it in the form of special 'low-fat' products, ballooned; and last but not least, trans-fats (fats artificially hydrogenated to make them stable at room temperature) proliferated. Each of these would have been bad enough on their own; together, they were a perfect storm.

Among the first to spot the dangers of the new fat phobia was the British nutritionist John Yudkin. As early as the 1950s, Yudkin had been warning that eating excessive carbohydrates – particularly refined sugars – could lead to spikes in insulin and obesity. Keys proceeded to attack Yudkin, pillorying him in the scientific press and dismissing his ideas as a 'mountain of nonsense'. To the detriment of millions, Keys won the debate: for another half-century, fat was off limits for most Americans and those, such as the British, who adopted their diet. An era of neurotic, cheerless eating ensued, as all foods with a trace of fat – chicken skin, egg yolks, whole milk – were condemned to waste. The food industry, meanwhile, went into overdrive, trying to find ways of making its fatless foods taste edible, the most obvious being to add more salt and sugar.[83]

As we now know, this period of joyless fat-free eating saw the steepest rise in obesity and diet-related disease in history – and all, it seems, for nothing. In 2010 the nutritionist Ronald M. Krauss published the results of a ten-year meta-analysis of all the evidence linking saturated fats to heart disease and concluded that there was none.[84] In 2014 a group of scientists supported by the British Heart Foundation reviewed seventy-two different studies and came to the same conclusion. Eaten in moderation, saturated fats were officially rehabilitated, while their usurper, sugar, was revealed to be just as *Pure, White and Deadly* as Yudkin had predicted in his book of that title back in 1972.[85]

Battle of the Bulge

> Nothing tastes as good as skinny feels.
>
> Kate Moss[86]

Today in the US and UK, dieting is a way of life. Each new fad is hailed as a miracle cure, and celebrities who shed the pounds are heralded as fat-busting idols, earning lucrative TV and book deals on the back of their shrinking figures. There is just one problem with all this: diets don't work. Oh yes, they may *seem* to at first: who among us has not leaped joyfully from the scales upon shedding a few pounds a week or two into some fancy new regime? The problem is that those pounds don't tend to stay off for long; often more pile back on than we shed in the first place. According to obesity expert Professor Jules Hirsch, this is because, regardless of how fat we are when we start a diet, our bodies react to dieting as though we are being starved. They take our starting weight as a set point and adjust our metabolism to recover the lost weight in any way they can.[87] A twenty-year study of dieters led by University of Minnesota psychology professor Traci Mann – the largest ever undertaken – found that the average weight loss over a two-to-five-year period was less than one kilogram. Furthermore, more than one third of dieters ended up gaining more weight than they had lost. Mann's conclusion? That we're better off not dieting at all.

If diets don't work, why do we keep going on them? One answer may be that the initial 'hit' we experience in the first few weeks makes us feel so good that we seek it again, giving dieting the quality of an addiction. This could help explain why those who diet tend to do so serially: one recent study found that the average forty-five-year-old British woman had attempted no fewer than sixty-one diets.[88] In Britain, the toll is particularly heavy among teenage girls, for whom celebrity culture and peer pressure – fuelled by social media – are proving a toxic mix. One recent survey found that only 33 per cent of fourteen- and fifteen-year-old girls in Britain felt good about themselves, with two thirds of those who didn't citing their looks – specifically feeling 'too fat' – as the reason why.[89] Fourteen per cent of those interviewed said they had skipped breakfast that day. The tragic endgame of such low self-esteem is also on the rise: from 2010 to 2018, hospital admissions in the UK due to serious eating disorders more than doubled from 7,260 to 16,023, while calls to the helpline of Beat, the UK's leading eating disorder charity, rose from 17,000 in 2017–18 to an estimated 30,000 in 2018–19.[90]

The fact that such disorders have increased with the rise of social media is no accident: Facebook, lest we forget, began as a game in which (male) college nerds could rate their fellow (female) students according to their 'hotness'. Today, thanks to Facebook and its imitators, being publicly rated by one's peers is an inescapable fact of life for teenagers. It is hardly surprising that, for many young people, eating, far from being joyous and carefree, is fraught with insecurity. The recent craze for so-called 'clean eating', in which celebrity influencers – many themselves past sufferers of eating disorders – push nutritionally dubious diets to credulous teens, is just the latest example of ways in which the subliminal link between food and sex can be used to prey on, and profit from, the vulnerable.[91]

Although ours is far from the first society to have had a perverse relationship with food (faddish diets have long been a social weapon), never before has food culture been so at odds with social aspiration. Instead of feeling nurtured by food, many teenagers feel they must fight it in order to gain social approval. For many, food has become the enemy.

Short of living on lettuce and running half-marathons, how can we thrive in an obesogenic world? Apart from the drastic cure of a gastric bypass, the only option for many seems to be to give up dieting altogether and simply accept that one is a tad lardy. A thirty-year study called *Eating Patterns in America* recently found that for the first time the number of dieting Americans was falling. In 1991, 31 per cent of adults surveyed said they were on a diet; by 2013, the figure was down to 20 per cent. The decline was greatest among women, down from 34 to 23 per cent.[92] A public revolt is under way in which the very idea of beauty is being reassessed. When asked whether they thought 'people who are not overweight are a lot more attractive' in 1985, 55 per cent of Americans agreed; by 2012, just 23 per cent did. Americans, it seems, are fed up with battling the bulge and feeling bad about themselves: from now on, it's official – fat is fabulous.

Just Add Water

> Worrying about something as simple as food in the digital age is weird.
>
> Rob Rhinehart [93]

Is affluence the enemy of joy? Epicurus certainly thought so, and his remedy was to seek happiness in non-material things. He spiced up his meals not with condiments but with conversation, an approach that modern neuroscience appears to endorse. Thanks to modern scanners, we know that chatting, playing games and reading books all light up our brains as reliably as eating a cupcake. Physical activity is also a dependable way of delivering dopamine to all the appropriate places: a brisk walk can improve our mood, and marathon runners experience a natural high after 'hitting the wall' about twenty miles in – as good an example as any of the complex relationship between pleasure and pain.

The advantage of seeking one's hedonic fixes from such pursuits is that one can do so almost without limits. Unlike eating, which soon makes us full, we can run, jump, sing, dance, ponder and puzzle from dawn to dusk if we want. Eating well is the basis of every good life, yet that includes being able to stop. Traditional food cultures around the world reflected this with periods of fasting. By contrast in industrial food nations, it is left to us to decide when not to eat, a choice that few of us are equipped to make.

Not everyone struggles with temptation, however; for some born in the digital age, eating is just a time-wasting adjunct to more exciting things. In 2012, a twenty-four-year-old engineer by the name of Rob Rhinehart was working in a San Francisco start-up designing phone masts. Feeling ill on his diet of fast food and frustrated by the time and money he was spending on it, Rhinehart decided to turn his engineer's brain to the question of how to eat. 'Everything is made of parts; every-thing can be broken down,' he wrote on his humorously titled blog, Mostly Harmless: 'I hypothesized that the body doesn't need food itself, merely the chemicals and elements it contains. So, I resolved to embark on an experiment. What if I consumed only the raw ingredients the body uses for energy?' [94]

By January 2013, Rhinehart had narrowed down the nutrients essential to healthy bodily function to around thirty-five ingredients, which he proceeded to order online. Top of his list were carbohydrates (sourced from oat flour and maltodextrin), lipids (from canola and fish oil) and protein (from rice), plus a cocktail of minerals and vitamins. Rhinehart threw his ingredients into a blender with water to create a beige goo (a non-slimming descendant of Metrecal) that he found tasted rather pleasant, a bit like pancake mix. He called his creation Soylent, after the Harry Harrison 1966 science fiction novel *Make Room! Make Room!* in which a post-apocalyptic New York City subsists on government rations of 'Soylent steaks' made from soy beans and lentils, which in the 1973 film *Soylent Green* were revealed to be made from human remains.

Rhinehart started living exclusively on Soylent, recording his experiences in a journal. At first not everything went to plan: on the third day Rhinehart found his heart racing and realised that he'd forgotten to add iron to his mix; a subsequent lack of sulphur caused him joint pain, and when he deliberately adjusted his levels of potassium and magnesium to see what would happen, he suffered cardiac arrhythmia and burning sensations. 'I wanted everything that could go wrong with Soylent to happen to me first,' he explained in a 2013 interview.[95] After one month, however, Rhinehart felt ready to share his experience with the world, writing a post on his blog entitled 'How I Stopped Eating Food':

> I haven't eaten a bite of food in 30 days, and it's changed my life . . . I feel like the six million dollar man. My physique has noticeably improved, my skin is clearer, my teeth whiter, my hair thicker and my dandruff gone . . . I sleep better, wake up more refreshed and alert and never feel drowsy during the day. I still drink coffee occasionally, but I no longer *need* it, which is nice.

The post went viral. Within days, Rhinehart saw that his future lay not in phone masts but in food. His jokey techno-sci-fi approach had instant appeal to science nerds for whom life hacking (short-cutting chores in order to concentrate on 'fun stuff') is a way of life. Rhinehart's willingness to share his formula soon created an eager DIY community keen to hack their own recipes and pool the results. By 2014, Soylent's site boasted more than 2,000 recipes from fifty-one nations with names such as People Chow, Hungry Hobo, I can't believe it's not food! and Grey Goo. Inevitably, hackers personalised their recipes by adding flavours such as vanilla and chocolate to make brownies, porridge and ice cream. In other words, they turned Soylent back into food.

Embracing his new-found celebrity status, Rhinehart used his blog to expand on his life philosophy. 'The world has changed,' he wrote. 'We don't live anything like our ancestors. We don't work like them, talk like them, think like them, travel like them, or fight like them. Why on earth would we want to eat like them? . . . In the past food was about survival. Now we can try to create something ideal.'[96] By 2015, Soylent was worth an estimated $100 million and Rhinehart had been approached by both NASA and the US military; in 2017, the drink went on sale in 7-Eleven stores.[97] To its legions of fans, Soylent really does appear to be a gastronomic philosopher's stone. The life hack of all life hacks, it seems to achieve what Rhinehart's hero Richard Buckminster Fuller called 'ephemeralization': the idea that technological progress means that humans will be able to do 'more and more with less and less until eventually you can do everything with nothing'.[98] Yet the nagging question remains: do we really want to do away with food, our greatest employer and most reliable source of pleasure?

One wonders what Epicurus would have made of Soylent. Would he have been an early adopter, cheerfully perfecting his personal formula and posting it online? It seems unlikely, yet in many ways his and Rhinehart's philosophies are uncannily close. Both men aim for a simple life; both are atomists who believe everything can be broken down into its constituent parts; both advocate self-discipline, and both appeal to reason to banish irrational fears – Rhinehart worked hard to overcome his fundamentalist Christian upbringing. Their key difference concerns the very thing that connects them the most, food. For Rhinehart eating and cooking are tedious chores that he happily consigns to oblivion, while for Epicurus they were the very heart of a good life.

Epicurus and Rhinehart hail from very different eras, yet their views transcend time. Had Epicurus lived today, he might have spent less time eating and more on Twitter, yet one suspects he would always have been up for a good natter and would still have baked his own bread. A good life comes in many forms – for each of us, it's a question of working out what brings us the most pleasure without harming ourselves or others. In this respect, Epicurus and Rhinehart represent opposite extremes, not just in their views on eating, but in their approach to life itself. While Rhinehart tries to escape necessity, Epicurus sought pleasure in satisfying it. Whatever you think of the two approaches, there is no question which is the more direct.

3
Home

A Finnish Farm

I am sitting at a long wooden table in a low, dark farmhouse. The house belongs to my Finnish cousin Helle. She is in her late fifties, but looks older to me, with her tanned, lined face and tired smile. Her farm, Rimpilä, lies in the heart of the Finnish lake district, about 200 kilometres north of Helsinki. My mother spent much of her wartime childhood here and has told us endless stories about life on the farm: how she used to love milking the cows and riding on the pigs' backs (they soon threw her off), and how she would take the huge farm horses to the lake for their evening swim. Now, for the first time, she has brought her English family here to meet their Finnish relatives. I am eight years old, tired from the journey and bewildered. This is the first time I've been abroad, and everything is new and unfamiliar. Nobody, apart from Helle's daughter Helena, speaks any English, and my mother is too busy catching up in the kitchen to bother to translate much of what is being said.

My father, brother and I sit silently at the table, which is beautifully laid with linen napkins and dainty plates of bread and butter, while the rat-a-tat of Finnish floats over our heads. My mother has explained to us that hospitality in Finland is taken very seriously, and we are conscious that we must do our best to eat whatever is served. We therefore await lunch with some trepidation. At last, Helle emerges from the kitchen carrying a round earthenware bowl containing some sort of brown pudding which she proceeds to serve each of us in turn. The pudding is unlike anything I've ever tasted: dense, smooth and incredibly rich, it is at once meaty, sweet and metallic, a combination that I am not at all sure I like. It is, we later learn, *maksalaatikko*, a dish made from calves' liver, syrup, eggs, rice flour, raisins and milk, traditionally served in Finland at Christmas or to special guests.

I dutifully eat my helping, grateful that my first ordeal is over. My father, however, doesn't get off so lightly. After accepting a large dollop and dispatching it with suitably appreciative noises, he is informed by my mother that Helle suspects that he doesn't really like the food and is just being polite, an observation that I know to be entirely accurate. Helle whisks the pudding away, only to appear from the kitchen moments later with an entirely fresh one, which she again offers to my father. With admirable fortitude, he accepts another large helping and resumes eating, but I can tell from his face that he is starting to flag. His is a Sisyphean task, however, since every time he clears his plate, another pudding appears, as if by some dread magic, in its place. After no fewer than five helpings my father, now looking decidedly wan, admits defeat; yet his efforts have not been in vain, since a delighted Helle at last concedes, through my mother, that he really must like the pudding after all, and is clearly a true English gentleman. Honour has been satisfied all round, and everyone starts to relax.

This ritual, which took place fifty years ago, belonged to a world that has now all but vanished. Although modern Finns remain extremely hospitable, few would stand and serve their guests as Helle did; fewer still would slaughter a calf in their honour, as we later discovered that our cousins had done. Most Finns today speak English, so the linguistic gulf that once separated the two halves of my family no longer exists. Back in the 1960s, however, in the absence of a shared language, my father and Helle resorted to a more ancient form of communication: the giving and receiving of food. Through the rituals of hospitality, they forged a bond more powerful than words.

My memories of that trip to Finland are full of vivid impressions. As a London child, I had never been on a farm before, so our visits to various cousins, most of whom were farmers, gave me my first taste of rural life. I loved it. I loved feeding the chickens and helping to bring in the cows for milking, exploring the forests and climbing over their ancient, mossy boulders. I loved the evening saunas we took in the woods, breathing in the heady pine-scented steam of their darkened interiors before cooling off in the silver lake, scarily close to a herd of wild elk.

I loved the hearty breakfasts of stew and buttery potatoes which I shared with several sunburned farmhands, for whom, I later learned, the meal was lunch.

It never occurred to me then that my cousins' way of life was unusual. I simply assumed that this was what country life was like. Yet already by that time, such 'Old MacDonald' farms had vanished from most of western Europe. In the 1960s Finnish farming remained largely in its pre-war state: most holdings were mixed, family-run affairs that still relied on horsepower. Due to their relative isolation, most were also largely self-sufficient. Recently, I asked my cousin Helena what it was like growing up on such a farm, and she told me about her 1940s childhood at Rimpilä:

There were about ten of us on the farm. There were five of us in the family, plus three women who helped in the house and dairy and two farmhands. We had five horses which we bred on the farm, and about fifteen cows for milk. We butchered the male calves to get good meat, and we also kept sheep, pigs and chickens. We grew rye, oats, barley, wheat and flax: with the flax we made yarn on handlooms, which we used to make our own sheets. We also knitted sweaters from our own wool. Twice a year in autumn and spring a seamstress would come to make clothes for everyone; a shoemaker also came to make us boots.

We had a big vegetable garden, where we grew potatoes, carrots, beetroot, cabbage, cauliflower, turnips, peas, onions, spinach, parsley, dill, lettuce, tomatoes and cucumber. We also had an orchard, where we grew apples, gooseberries, rhubarb, blackberries, plums and cherries. Our gardener was very clever: he knew how to make plums grow, even that far north! We went fishing in the lake, and we foraged for nettles, mushrooms, berries and sorrel. My mum also used to gather wild raspberry leaves to make tea. In summertime we had plenty of things to eat! It was a lot of hard work, but we had enough farmhands, so that was fine. We made lots of jams and preserves, and we also had stores in a separate building for meat, butter and flour. Of course we made our own bread. We took our grain to the local mill, which was a water mill, about five kilometres away.

We didn't need to buy much, and for that we sold our milk to the local dairy and our sheep and wool at market. We also sold our timber: in many ways, we lived off the forest. My mother used to buy lamp oil, matches, tea, salt, sugar and coffee if she could get it; it was considered very impolite not to serve coffee to people if they came to visit. I think my mum also bought wheat flour if she wanted to make *pulla* [a sweet Finnish bread]. The war made it difficult to import things, so it was like this until the early 50s, but soon afterwards it began to change. We began to import rice, so we had rice porridge for Christmas after the war. Before that we made porridge with barley grains, which was actually tastier but took longer to cook.

Listening to Helena, I marvel at the fact that her childhood, which took place just twenty years before mine, could have been so very different. I admire the resilience and skill that she and her family displayed in living as they did. The Finns have a word, *sisu*, which means strength, grit and tenacity in the face of adversity. It's a quality, you might say, that one needs to survive in a place where it's cold and dark for half the year and the next village is twenty miles away. Such fortitude and the Finns' notorious capacity to stay silent for hours on end are no doubt related. You can't take the geography out of the person: we're all shaped by the place where we are born.

It's easy to idealise such a way of life, so I ask Helena whether she remembers her childhood with fondness. Yes, she says, life at Rimpilä was good: it was physically tough but also felt friendly and full of a sense of purpose. Reflecting a little, she adds that, since she didn't know any other way of life, she readily accepted hers. Despite the war, she says, her childhood was happy.

Rimpilä is now run by Helena's nephew Kalle. With the help of modern machinery, he manages mostly on his own, employing temporary labour when necessary. He grows commercial crops – wheat, barley, oats and rape – and sells timber. He also keeps some land as meadow, in order to preserve some natural diversity on the farm. The farmhouse today is as neat as a holiday cottage: there are no chickens running about in the yard, only cars, and the old wooden barn where my mother once slept above the cows is empty. The village hall up the road where

Helena once danced with local young farmers is shut. Once thronging with people, animals, lives and loves, this part of rural Finland is now mechanised and depopulated, a place where survival depends on growing cash crops as efficiently as geography, technology and conscience will allow.

Rimpilä's story is a familiar one. It mirrors the march of modernity across the world – the latest chapter in a centuries-old saga in which people have gradually left the land in order to live in cities. It tells of our unfolding relationship with technology and our use of it to ease the burden of feeding ourselves. The shift from rural to urban life has changed our lives, many would say for the better. Yet at what cost has our freedom come? Our need to eat once located us: it told us where we belonged. Now that link is broken, how do we find ourselves in the world? How, in particular, do we find that special place we call 'home'?

A Place of One's Own

> . . . our house is our corner of the world.
> Gaston Bachelard[1]

What and where is home? In its most straightforward sense, it's the building or place where we live: the house, flat, cottage or caravan where we keep our stuff and return each night to sleep. Yet 'home' has other meanings too: it might refer to the place where we were born, the region or neighbourhood where we grew up or some distant ancestral land we've never seen. Wherever it is, however, its deepest meaning will always be the place where we feel that we most belong.

On the grand scale, of course, home for all of us is planet earth. Home for orbiting astronauts is not their spaceship, but the planet they can see out of its window. The yearning for home that many report feeling at this sight leaves them permanently transformed. Even for those of us who remain earthbound, the image of earth from space – as immortalised by the famous 1972 'Blue Marble' photograph, taken by the crew of Apollo 17 – is deeply resonant, so much so that it has been credited with changing the way we humans relate to our planet.

As such powerful responses suggest, home is both a place and an *idea*. In order to feel grounded, we must be able to orientate ourselves in the world. Home, in this sense, is our anchor, a place from which we might wander, yet around which our emotional lives will always revolve. Yet there is nothing fixed about home: it can be mobile or static and can exist on any scale, from a shack or a ship to a house, village, city, landscape, nation or planet. One thing that home *can't* be, however, is somewhere we can't eat. Home must be able to sustain life.

In this sense, our first home is our mother. Before and immediately after we're born, our mothers give us food, love and protection all at once: the three pillars upon which all future home life will be based. The tastes of our childhood are etched into our psyches for this reason: they come from the time when we learned what it means to feel at home. Whatever our childhood home is like, it will be the most formative we'll ever have. As Gaston Bachelard wrote, 'The house we were born in is physically inscribed in us.'[2] Our first home is also where we learn to behave as social beings, not least around the family table.

Just as we have to learn how to eat, we have to learn how to share, making the family meal our most significant early training in how to behave as good political animals. We may not realise it, but the shared meal remains the most primal ritual that we commonly perform. As adults, we take for granted that those we eat with won't steal our food – apart, perhaps, from nicking the odd chip. Yet to watch other omnivores sharing food is to realise quite how precious this assumption is: after a hunt, for example, male chimpanzees begin a frenzied fight over the spoils, tearing at the meat and uttering screams that can be heard up to a kilometre away.[3] Weaker males and females tend to stick to vegetables.

Alone among species, we've evolved ways of sharing food without the threat of violence. That's not to say that it is not latent whenever we eat together; on the contrary, table manners remain a vital and underestimated survival skill. If they seem overly concerned with how to fold one's napkin or hold a soup spoon, it's only because deeper behavioural codes are so engrained in us as to be imperceptible. Like the act of eating itself, however, such codes are far from innate: young children sharing cake at a tea party, for example, can resemble those crazed chimps. Learning to share food well is critical to our upbringing, not just so that we don't steal all the cake, but because it teaches us the self-restraint and mutual trust necessary to live as civilised beings.[4]

The significance of sharing food is reflected in our language: a companion, for instance, is both someone with whom we break bread (from Latin *com*, with + *panes*, bread) as well as the sort of person we might trust enough to form a company. As Margaret Visser noted in *The Rituals of Dinner*, hospitality brings us together, turning 'host' and 'guest' as well as potentially 'hostile' outsiders (all from the Indo-European *ghostis*, stranger) into friends.[5] Eating is above all a social activity. We find eating alone deeply unnatural; indeed, the anxiety we often feel when dining alone in public stems from an ancient instinct, as our wild ancestors were most vulnerable to attack when they were eating. When sharing food with those close to us, by contrast, we feel happy, safe and relaxed. Such meals reinforce our sense of belonging to a tribe as few other rituals can.

The power of food to bind us together has long been recognised and widely exploited. As Jean Anthelme Brillat-Savarin observed, 'there has never been a great event, not even excepting conspiracies, which was not conceived, worked out, and organised over a meal'.[6] Yet only recently have we started to understand how bonhomie and biology fuse. As the British anthropologist Robin Dunbar explains, the act of eating together triggers endorphins – opioids chemically related to morphine that form part of our brain's pain management system – that give us a natural high. Sharing a meal thus has a similar effect to affectionate gestures such as stroking and cuddling, helping us to bond, like our primate cousins, in social groups.[7]

As the US neuroscientist Paul J. Zak discovered, sharing meals also triggers oxytocin, a hormone involved in breastfeeding and other forms of emotional bonding. Zak stumbled across the effect when trying to work out why subjects in economic 'trust' experiments (involving exchanges of cash between participants) behaved more generously than the models predicted. When participants behaved magnanimously towards their partners, Zak found, the latter experienced a surge of oxytocin, prompting them to respond in kind.[8] Dubbed the 'moral molecule' by Zak, oxytocin prompts us to obey the golden rule 'Do unto others as you would have them do unto you.' It helps us to trust, in other words, that if we refrain from helping ourselves to a large slice of pie, others will do the same and we'll all end up with seconds.

Terra Cognita

Conviviality affects every aspect of human life.
Jean Anthelme Brillat-Savarin[9]

Food makes us feel at home because it roots us in the world, both socially and physically. In the past such rootedness was literal, since our ancestors lived almost exclusively on what they could grow or gather locally. Today, however, most of us live in cities, far from the land that feeds us. So how does that affect our sense of home?

The answer is that, wherever we live, food remains central to our sense of identity and belonging. A full English breakfast may consist of Danish bacon, Dutch sausages and French eggs, yet the thought of a good old fry-up is still familiar and comforting for many Brits. Does it matter that the foods we identify with are, in some respects, a fiction? Do we do ourselves some sort of existential damage by sipping a Frappuccino in a fake-Seattle Starbucks? Whatever the answer, there is no doubt that we are hard-wired to respond to foods that feel authentic to us because they remind us of home, even when we're far away.

Every Sunday in Hong Kong, thousands of Filipino maids gather to share a picnic, congregating in the city's central business district known as Central, occupying such unlikely spots as the undercroft of the Hong Kong and Shanghai Bank. Under these glinting corporate facades, the women spread out rugs and take out Tupperware boxes of spicy foods to share while they gossip, sing songs and read letters from families and friends, eating with their hands as they would at home. Women from different regions gather in specific locations, transforming Central into a gustatory map of the Philippines whose incongruous odours can irritate the neighbours, yet which make the women feel more at home than they do at any other time of the week.[10]

Migration and exile have always been part of the human experience, making the lost or abandoned home – as exemplified by the Garden of Eden – a key mythical theme. The trope reveals a deep tension in our lives between the need to seek our fortune in the world and our yearning for what we must leave behind. Migrants can feel such nostalgia for home that they try to replicate it in their adopted one, a habit that can have unintended consequences. Early Swedish settlers in North America, for example, replicated the farms and villages of home by building log cabins in the woods. Such dwellings are now considered archetypically American, usurping the portable tepees and grass wigwams in which Native Americans actually lived.[11] For them home was never a building, but rather a carefully curated territory that, because of its lack of farms, fields and fences, went unrecognised by Europeans.

Home is a response to landscape formed by an idea of how to live. It is always shaped by food: if one lives by gathering berries and hunting bison, for example, one's home is going to look very different to that of someone who farms. We inhabit the world according to our inherited culture, the cumulative result of generations of our ancestors' efforts to survive in a certain terrain. Over time, humans have settled in jungles, deserts, grasslands, forests, mountains and on coasts, places that, although wildly different, have all yielded enough food for us to survive. By making such wildernesses habitable – by finding sites, lighting fires, fending off beasts, drawing water, gathering fuel and finding things to eat – we have made ourselves at home. By cooking meals and building shelters, we have created camps, farms, villages and eventually cities. By working out how to feed ourselves, we have found our place in the world.

Life in the Larder

In the past, the direct link between food and home was recognised through foundation rites that symbolically bound the two together. In India, for example, a *gharbha*, a sacrificial vessel, was filled with earthly treasures such as gems, soil, roots and herbs and placed in the foundations of a future temple or city.[12] The Romans also founded cities on a sacred pit, a *mundus*, into which offerings of food to the gods of the underworld were placed in the hope that well-fed divinities would bless the soil and make it fertile, so that the city would flourish.[13]

Both farmers and city dwellers in the pre-industrial world felt a powerful connection with the land that fed them, yet our hunter-gatherer ancestors felt a stronger connection still. For them, home was, quite literally, the larder, as the English anthropologist Colin Turnbull discovered when he spent three years with the Mbuti pygmies of the eastern Congo in the 1950s. In his classic book *The Forest People*, Turnbull described a way of life that had barely changed for millennia. Living deep in the forest, the Mbuti relied on timeless knowledge and skills to survive and were utterly at one with their habitat:

> Whereas the other tribes are relatively recent arrivals, the Pygmies have been in the forest for many thousands of years. It is their world, and in return for their affection and trust it supplies them with all their needs. They do not have to cut the forest down to build plantations, for they know how to hunt the game of the region and gather the wild fruits that grow in abundance there, though hidden to outsiders. They know how to distinguish the innocent-looking itaba vine from the many others it resembles so closely, and they know how to follow it until it leads them to a cache of nutritious, sweet-tasting roots. They know the tiny sounds that tell where the bees have hidden their honey; they recognize the kind of weather that brings a multitude of different kinds of mushrooms springing to the surface; and they know what kinds of wood and leaves often disguise this food. The exact moment when termites swarm, at which they must be caught to provide an important delicacy, is a mystery to any but the people of the forest. They know the secret language that is denied all outsiders and without which life in the forest is an impossibility.[14]

Their deep familiarity with their habitat made the Mbuti totally fearless and relaxed, Turnbull found. They referred to the forest as 'Mother' or 'Father' since, as one elder explained, 'The forest is a father and mother to us, and like a father or mother it gives us everything we need: food, clothing, shelter, warmth and affection.'[15] The Mbuti seemed content with life, Turnbull noted, frequently smiling or laughing. 'They were a people,' he wrote, 'who had found in the forest something that made their life more than just worth living . . . something that made it, with all its hardships and problems and tragedies, a wonderful thing full of joy and happiness and free of care.'[16]

The sense of feeling profoundly connected to their territory appears universal among hunter-gatherers. In a similarly famous 1962 study of the western Australian Walbiri, *Desert People*, the Australian anthropologist Mervyn Meggitt noted how the Walbiri viewed the landscape as a living embodiment of their ancestors, whom they believed to have created it in the past time known as the Dreaming. Young boys being prepared for initiation into adulthood would be taken on a tour of the surrounding landscape by a guardian and an older relative that lasted two to three months, during which time they were introduced to the local flora and fauna, learning their practical and totemic significance. Every form in the landscape had a story of how it came to be made: a rocky outcrop, for instance, might echo the form of a sleeping ancestor; a waterhole could be the place from which another might spring.[17]

As the British social anthropologist Tim Ingold has argued, this 'show-and-tell' form of teaching instils a particular kind of knowledge in the novice: rather than just cramming his mind with facts, it gives him a sense of ancestry and belonging. In such an 'education of attention', the boys gain not just information but awareness: 'Placed in specific situations, novices are instructed to feel this, taste that, or watch out for the other thing. Through this fine-tuning of perceptual skills, meanings immanent in the environment . . . are not so much constructed, as discovered.'[18] The resultant feeling that hunter-gatherers have for their environment, Ingold suggests, is not unlike that which potters might feel for clay or joiners for wood. The sense of material connection for them is omnipresent and palpable: 'a mode of active, perceptual engagement, a way of being literally "in touch" with the world'.[19] For our hunter-gatherer ancestors, home was more than a place to live; it was a territory whose every feature was familiar and alive and full of purpose, connection and meaning.

Hearth and Home

> The oldest of all societies, and the only natural one, is that of the family.
>
> Jean-Jacques Rousseau[20]

Wilderness was home to our ancestors, yet once they learned to control fire, the hearth became its core.[21] Our earliest cooking forebears, *Homo erectus*, lived in bands of thirty to sixty members without a formal leader, and the hearth was the focus around which life revolved. Most of the men would leave camp each day to go hunting or foraging, while the women stayed behind to tend the fire, look after the children and hunt for tubers to cook an evening meal that could feed the whole group, should the hunt fail. As Richard Wrangham has argued, cooking such a reserve meal was crucial to the band's success, since it enabled some members to become specialist hunters, to the benefit of everyone.[22]

'You hunt, I cook' is arguably the oldest form of social contract. The exchange of one sort of food for another – meat or honey, say, for cooked roots – remains unique to humans. In effect, this division of labour created a mini-economy, a bond of trust that formed the nucleus of a new social institution, the family household.[23] From the first, the arrangement was split along gender lines: since men are stronger and faster runners than women, it made sense for them to hunt, while the women stayed in and around camp to attend to domestic tasks. The agreement was reflected in the way that food was shared: just as hunter-gatherers do today, the returning hunters would almost certainly have shared the best spoils with their partners and children, before passing the rest of the food to other members of the band. In this way, the family became the primary social unit within the wider group, an arrangement that has proved remarkably enduring.

As Edward O. Wilson observes in *The Social Conquest of Earth*, living with fire would have required social skills beyond any that had gone before. Communication, cooperation, trust and empathy would have been necessary in order for the new divisions of labour and reward to function. Of these, the capacity to empathise would have been paramount, since reading others' intentions and forming alliances would have been essential to survival in the competitive environment of the camp.[24] Stone-Age levels of oxytocin must have been sky-high.

Over time, the complexities of camp life fostered humanity's greatest invention, language. This made more complex social interaction possible, including the shared rearing of children, as well as giving rein to a range of behaviours from generosity to duplicity. Since any individual's success depended on their status in camp as well as that of the group as a whole, a tension existed between selfishness and altruism. The result, says Wilson, would have been 'highly flexible alliances, not just between family members, but between families, genders, classes and tribes'.[25] To succeed in such a febrile milieu would have required the full gamut of social skills, not all of which were honourable. 'The strategies of this game,' says Wilson, 'were written as a complicated mix of closely calibrated altruism, cooperation, competition, domination, reciprocity, defection and deceit.'[26]

Our ancestors, it appears, would have felt quite at home in the Palace of Westminster. Indeed, many scholars now believe that the feverish politics of camp life fostered the 'cognitive revolution' that took place 70,000–30,000 years ago, when *Homo sapiens* entered a phase of creative turbocharge, inventing such things as boats, oil lamps, bows and arrows and that essential accoutrement of civilised society, art.[27]

Cultus

> It is iron and corn, which have civilised men, and ruined mankind.
> Jean-Jacques Rousseau[28]

Cooking wasn't the only thing that our ancestors did with fire. Around 45,000 years ago, they began using it to burn forest and scrub – as Aboriginal Australians still do today – in order to encourage new plant growth and capture game.[29] Humans, that is to say, began to shape the landscape and domesticate animals – practices that around 12,000 years ago evolved into a radical new way of feeding themselves: agriculture.

It is hard to overstate the impact of farming on human history. It changed just about everything our ancestors did: what they ate, how they lived and how they saw the world. Although humans had eaten seeds for millennia, the habit of cultivating them – not only planting them, but sticking around to harvest and choose the best seeds for the next year's planting – was revolutionary. Cereals are rich in energy and can be stored in bulk, meaning that for the first time a small piece of land (a field) could yield enough food to feed those working on it throughout the year. From now on, there was no need to keep moving; people could settle and spend more of their time on things other than hunting and gathering: poetry, pottery and counting ensued, followed by architecture, institutions and eventually cities.[30]

The link between farming and civilisation wasn't lost on our ancestors. The Roman word *cultus*, for example, meant both cultivation and culture. Mesopotamians, Mexicans, Greeks and Chinese all agreed that eating grain had civilised mankind, their creation myths variously celebrating the discovery of wheat, maize or rice.[31] Not everyone appreciated the change, however: Genesis is far from the only ancient text to portray farming as a divine punishment. Early farming was much harder work than hunting and gathering had been; indeed, the very concept of work only appeared when people started to farm.[32] Among contemporary hunter-gatherers, who may spend as little as twenty hours a week actively hunting or foraging and for whom such tasks are embedded in everyday sociability and ritual, the concept is virtually unknown.[33] For such people work is simply life.

Life for early farmers, by contrast, must have felt like nothing but work. First they had to clear and plough their land, then sow and water their crops and protect them from pests and predators, all the while staring heavenwards in fear of too much or too little sun or rain. The grain had to be harvested at precisely the right moment and the edible germ parted from its husk, a laborious task that may have involved toasting the grains before rolling them between stones to create a rough flour. This was combined with water to make a gritty dough or porridge; it was only around 5000 BCE that someone had the bright idea of baking the dough into the staple that still feeds one in three of us today, bread.

For all its sweat and toil, farming didn't deliver better health; on the contrary, early farmers were smaller than their hunter-gatherer counterparts and lived shorter lives. The fossil remains of hunter-gatherers in

Greece and Turkey around the end of the last Ice Age, for example, reveal an average height of 5 feet 9 inches for men and 5 feet 5 for women; by 3000 BCE this had fallen to just 5 feet 5 inches and 5 feet.[34] Records from the Illinois River Valley also suggest that life expectancy dropped from twenty-six to nineteen when people started to farm.[35] Thanks to their newly restricted diet, early farmers also suffered from a range of previously unknown conditions such as rickets, scurvy and anaemia.[36]

The Neolithic Housewife

If farming was so unrewarding, why did people bother with it? The short answer is that they had no choice. As the last Ice Age ended and the climate began to warm, the happy hunting grounds of the real-life Garden of Eden – the verdant forests of the so-called Fertile Crescent – receded northwards, leaving grasslands and much smaller game behind. As the population increased thanks to the warmer, drier climate, it was a case of adapt or die.

Popular or not, farming certainly spread rapidly. From its origins in the Fertile Crescent around 10,000 BCE, it had spread by 5000 BCE to every continent apart from Australia, and by 2000 BCE most humans were farmers.[37] By this time, most of the plants and animals we still rely on today – wheat, maize, rice, barley, rye, cows, pigs, chickens, ducks and goats – had been domesticated. Despite the parallel rise of urban life, the vast majority of people still lived in rural areas – as they did until the twenty-first century. In 1800, just 3 per cent of the global population lived in settlements of 5,000 inhabitants or more; in 1950 the figure was still less than one third.[38] Although most of us now live in cities, home for most people in history has been a farm.

Farming's most fateful effect was arguably to alter our relationship with nature. Wilderness, once home to every human, became the enemy. Agrarian myths spoke not of benign ancestors dormant in the landscape, but of vengeful gods who might withhold the necessary sunshine, rain or fertility.[39] Home changed too: no longer a sanctuary within nature, it became a territory wrested from it, transformed through hard work. As people began living in buildings, home became indoors instead of outside, warm rather than cold, private, not public. Home, in short, became a realm set apart from the world: a household in the true sense.

There was, however, one aspect of domesticity that survived the transition: the division of labour between men and women. While Neolithic menfolk ploughed fields, felled trees and built fences, women stayed closer to home, tending and harvesting crops, grinding grain to make flour, looking after children and animals, and of course cooking. The complaint of 1970s housewives that domestic arrangements remained mired in the Stone Age wasn't far wrong. The ancient pact between hunter and cook has lasted until our own day – a deal in which those who have stayed indoors have been as consistently undervalued as they have been vital to the success of our species.

Home Economics

> A man acquires nothing better than a good wife, and nothing worse than a bad one.
>
> Hesiod[40]

In ancient Athens, houses themselves reflected this patriarchal pact. Most were arranged around open courtyards where the family lived and cooking was done, with a more formal, public area facing the street. The most important space here was the *andron* (man's room), a dining chamber in which important guests were entertained to *symposia*, networking dinners around which social and political life in the city revolved. As the nomenclature makes clear, public life in Athens was an all-male affair, and the private realm of women, children and slaves was *idion*, from which our word idiot is derived.

For Aristotle, the household (*oikos*) was the bedrock of the *polis*. As he explained in his *Politics*, households were naturally formed when men and women came together in order to reproduce.[41] Having set up house together, the couple's next step was to find a way of feeding themselves and their family, a process as natural to humans as it was for animals. Such household management, or *oikonomia* (from *oikos*, household +

nemein, manage), underpinned the state, since it formed the basis of the self-sufficiency that was, for Aristotle, its 'end and perfection'.[42]

Oikonomia, as you may have guessed, is the root of our word economy – which incidentally makes the term home economics somewhat tautological. For the Greeks, however, *oikonomia* was about a great deal more than baking cakes; it was the foundation of a good life, as the poet Hesiod made clear to his younger brother in his eighth-century-BCE tract *Works and Days*:

> I suggest you reflect on the clearing of your debts and the avoidance of famine. First, a household, a woman, and a ploughing ox – a chattel woman, not wedded, one who could follow the herds. The utilities in the house must all be got ready, lest you ask another, and he refuse, and you be lacking, and the right time go past, and your cultivation suffer. Do not put things off till tomorrow and the next day. A man of ineffectual labour, a postponer, does not fill his granary.[43]

Hesiod's lecture, which continues at some length, is effectively the world's first agronomic treatise. It describes how the farmer should run his household like a business, an enterprise in which his wife is expected to act as a docile factotum ('chattel', tellingly, shares a root with 'cattle'), not as an equal partner with her husband, who is naturally also in charge of the animal 'husbandry'.

The detailed description of Greek home life to be found in Xenophon's fourth-century-BCE *Oeconomicus* suggests that such arrangements remained firmly entrenched in democratic Athens. The *Oeconomicus* follows Xenophon's teacher Socrates as he quizzes the model citizen and farmer Ischomachus on the running of his household. Typical of a wealthy Athenian of the time, Ischomachus owned a large house in the city and a similarly extensive farm in the countryside, with plenty of slaves to work both. Ischomachus begins by telling Socrates how he chose his wife at the age of fifteen for her high level of self-command and discipline, and how he instructed her on the running of his household: 'Your duty will be to stay indoors and dispatch those slaves whose work is outside, and receive what comes in and dispense as much of it as must be spent, and watch over as much as is to be kept in reserve, and take care that the amount stored up for a year is not spent in a month.'[44]

Ischomachus tells his wife that she will be like a 'queen bee in the hive', while he rides out each morning to the estate in order to supervise whatever work may be under way, 'whether it be planting, clearing, sowing or harvesting'. After putting his horse through its paces, he hands it over to a slave to be brought back home with any provisions needed for the house. Ischomachus himself walks or runs back home for lunch, before spending his afternoons on business in the city. 'I certainly do not pass my time indoors,' he tells Socrates, 'for my wife is quite capable of managing the household, even by herself.'[45] As for his wife's health, Ischomachus advises her to get stuck into the housework in order to stay fit: 'I also said it was excellent exercise to mix flour and knead dough; and to shake and fold clothing and linens; such exercise would give her a better appetite, improve her health, and add natural colour to her complexion.'[46]

Socrates professes himself delighted with these arrangements, which were very much in line with the Greek ideal of a good life. The importance of *oikonomia* was not just due to the idea of economic self-sufficiency, but because of its opposition to *chrematistike*, the pursuit of wealth for its own sake. Unlike *oikonomia*, which had natural limits, the pursuit of wealth could never be satisfied, and thus could never bring happiness. As Aristotle argued, those who pursued *chrematistike* were 'eager for life, but not for the good life'.[47]

Family Fortunes

> The world we have lost, as I have chosen to call it, was no paradise.
>
> Peter Laslett[48]

While large, complex households of the sort described by Xenophon persisted in southern and eastern Europe and remain common in the Middle East, another type of household emerged from about 1500 in northern Europe, as women began postponing marriage until their late twenties, spending their early adult years working away from home.[49] The change was partly due to new patterns of land tenure after the Black Death, when many serfs were freed to become tenant farmers on their own land.[50] Since the plots were small, children were often forced to leave home to find work elsewhere, either until they could inherit their parents' farm or marry and buy one of their own.[51]

Late marriage put men and women on a far more equal footing. Rather than enter their husband's family as a teenage bride, wives set up independent households with their husbands and often ran them as equal business partners.[52] Women were increasingly valued, not for their diligence in folding linens, but for their earning capacity – a change reflected in the fact that many remarried within a year of being widowed.[53] On top of childcare and cooking, rural housewives made butter, cheese, bread, beer and preserves, kept poultry, took in washing, spinning or sewing and kept the books, as well as dealing with traders and suppliers. As the historian Judith Flanders has noted, one doesn't have to puzzle over the origins of the proverb 'A woman's work is never done.'[54] Despite the hard work, however, northern European wives bore fewer children than their Mediterranean counterparts, and so lived longer and in greater numbers.

Life in urban working households was similarly hard. One 1619 London bakery consisted of twelve or thirteen people in total: a master baker and his wife plus two or three children, four journeymen, two apprentices and two maidservants.[55] There was a workshop on the ground floor, with stores for grain, sea-coal and salt out the back and living quarters above. Everyone ate together and, apart from the journeymen (skilled workers who journeyed from one master to the next and thus had a degree of independence) slept under one roof: the householders lived, that is to say, as one large family.

Childhood in such pre-industrial households was tough. Since many children died before adulthood and were needed for their labour, there was little incentive to invest much emotion in them. Apprentices as young as ten were covenanted to live with a chosen master for up to seven years, and the master's own children were put to work from an early age: John Locke declared in 1697 that all should work from the age of three.[56] Many children were sent away to work in service, sometimes swapped by their parents for someone else's, presumably on the basis that other people's offspring were more biddable than one's own. In 1800, 40 per cent of northern Europeans had been servants at some point in their lives; among women the figure was as high as 90 per cent.[57] As the historian Peter Laslett remarked, 'The coming of industry cannot be shown to have brought economic oppression and exploitation along with it. They were there already.'[58]

Farm and Factory

And was Jerusalem builded here / Among these dark Satanic Mills?
William Blake[59]

As the descriptions in the last section suggest, our idea of family life today is light years away from its pre-industrial counterpart. Before 1800, a 'family' was simply a group of people who happened to live and work under one roof. Far from being a place of rest, the household was the mainstay of the economy in both city and country, its primary function to foster not comfort and love but hard-nosed productivity. By modern standards, such houses were far from cosy: rural homes in particular were often dark, damp and cold, and most contained little furniture, but rather the tools and materials needed for work. Although the wealthiest in society enjoyed far greater comfort, life for the majority was unambiguously harsh. Why, then, do we persist in feeling nostalgic about life before the age of steam?

The answer is partly because what followed was, at least to begin with, so much worse. When cotton mills first appeared in Lancashire during the 1720s, it spelled the beginning of the end of traditional British rural life, an existence which, although tough, had fostered strong connections between people and land. Village life demanded constant cooperation, particularly with seasonal tasks such as bringing in the harvest. In this sense, villages were an extension of the household, a collective space in which activities such as fetching water, going to the mill or harvesting crops replicated some of the ancient sociability of the tribe.

As the Industrial Revolution gathered momentum from the 1770s onwards, all this was swept away, as land was forcibly enclosed by Parliament in order to boost agricultural production to feed the rapidly expanding cities. Between 1761 and 1844, four million acres of village-based open strip fields were transformed into the neatly hedged, privately owned rectangles we know today. Over the same period, two million acres of commons and wasteland (public woods, marshes and moors free for all to use) were brought under cultivation.[60] Such enclosures not only deprived peasants of land to farm, but robbed them of the commons that had formed a vital part of the rural economy. Constituting one third of all land in England and Wales in 1688, commons were used to graze livestock, to catch a bird or rabbit for the pot and to gather materials such as wood or peat for furniture and fuel, moss and bracken for bedding, as well as rushes, wax, honey, wild plants, berries and herbs.[61] They were, in short, the shared resource that made *oikonomia* possible. Without them country life was cut off at its roots. Widespread poverty, worsened by the agricultural depression that followed the Napoleonic Wars, created a stricken population.[62] For many, working in factories was the only option.

The poet Oliver Goldsmith was just one of many commentators who lamented the vanishing rural order. In his 1770 poem *The Deserted Village*, Goldsmith evoked a mood of nostalgia for a disappearing world:

> Ill fares the land, to hasting ills a prey,
> Where wealth accumulates, and men decay:
> Princes or lords may flourish, or may fade;
> A breath may make them, as a breath has made;
> But a bold peasantry, the country's pride,
> When once destroyed, can never be supply'd.
>
> A time there was, ere England's griefs began,
> When every rood of ground maintained its man;
> For him light labour spread her wholesome store,
> Just gave what life required, but gave no more:
> His best companions, innocence and health;
> And his best riches, ignorance of wealth.

Not everyone, however, lamented the passing of the peasantry. In his 1783 poem *The Village*, George Crabbe wrote a riposte to Goldsmith and those like him who romanticised the harsh realities of rural life:

> Yes, thus the Muses sing of happy swains,
> Because the Muses never knew their pains:
> They boast their peasant's pipes; but peasants now
> Resign their pipes and plod behind the plough;
> And few, amid the rural tribe, have time
> To number syllables and play with rhyme.[63]

The social upheaval that accompanied the Industrial Revolution and ensuing tensions over whether or not it was ultimately a good thing began a debate that remains as pertinent today across the world as it was in eighteenth-century Britain. Whatever one's view of industrialisation, there is no question that it put an end to an ancient way of life. For all their downsides, pre-industrial working households had delivered benefits that factory labour could never replace: collaboration, companionship, craft, knowledge and skill, and at least the chance of economic independence.[64] Factories offered none of these things. What they demanded was a cheap, low-skilled workforce that could be hired and fired at will. In place of the intimate bond between master and apprentice – oppressive at worst, yet transcendent at best – came the certainty of anonymous exploitation.

In 1842, a twenty-two-year-old German by the name of Friedrich Engels arrived in the English industrial capital, Manchester. Engels was so shocked by what he found, not least in the notorious back-to-back terraces of the Old Town where many factory workers lived, that he felt moved to write a tract entitled *The Condition of the Working Class in England*:

> It is far from black enough to convey a true impression of the filth, ruin, and uninhabitableness, the defiance of all considerations of cleanliness, ventilation and health which characterise the construction of this single district . . . If anyone wishes to see in how little space a human being can move, how little air – and *such* air! – he can

breathe, how little of civilisation he may share and yet live, it is only necessary to travel hither.[65]

Extreme poverty wasn't the only outcome of factory work; another was the shift of power between the sexes. As power looms were introduced in mills, jobs that required physical strength or weaving skills – the traditional preserve of men – went into steep decline. Women and children were both cheaper and had nimbler fingers better suited to tending the new machines. By 1834, just one quarter of mill workers were adult men, with the rest of the workforce consisting of women and children.[66] The role reversal wrought havoc with family life, as Engels noted, with men forced to stay at home and do 'women's work':

> In many cases the family is not wholly dissolved by the employment of the wife, but turned upside down. The wife supports the family, the husband sits at home, tends the children, sweeps the room and cooks . . . it is easy to imagine the wrath aroused among the working men by this reversal of all relations within the family, while the other social conditions remain unchanged.[67]

'Can anyone imagine a more insane state of things?' Engels demanded. His outrage would soon be deployed in the *Communist Manifesto*, in which he and Karl Marx would dissect a new economic order that reckoned human value, not by skill, ability or character, but by a single commodity, labour.

Home Sweet Home

> Home . . . is the place of Peace.
> John Ruskin[68]

Just as farming had done before it, industrialisation transformed the nature of home. Households were the productive engines of the pre-industrial economy – in that sense, little had changed since the days of the ancient Greeks. With the advent of factories, however, work, newly defined as labour rewarded by wages, moved outside the home. Instead of producing their own food, furniture and clothes, people would henceforth earn the cash to buy them ready-made. From centres of production, homes became places of consumption, a role that would prove crucial to the industrial project.

By the mid-nineteenth century, industrialisation was starting to pay dividends. The Factory Acts of 1833 and 1847 limited the hours that women and children could work, while men could aspire to a new raft of jobs in business, management and the professions that earned enough for their families *not* to have to work: the definition of the emergent middle class, which by 1851 represented around 15 per cent of the popu-lation.[69] Children started to be seen in a new light: while the poorest still contributed up to half of a family's income, middle-class children became little treasures to be educated, cosseted and adored. As the social historian Theodore Zeldin notes, 'The role of children became to spend their parents' money, instead of earning it.'[70] Women, meanwhile, took on the role of full-time wives and mothers, replicating the ancient realm of *idion*.

Domestic tasks were reclassified as 'housework', a form of labour which, by virtue of being cut off from the income stream – like an eco-nomic oxbow lake – was no longer valued. In the industrial economy, only labour rewarded with wages would be considered work. Once integral to the productivity of the household, housework was relegated to the thankless, lonely chore that it remains today. Women whose grandmothers had once made bread and put it on the table now had to wait for their breadwinner husbands to earn enough to buy it.

The arrival of the railways only served to make such divisions starker. Victorian middle-class workers were the first to commute, creating a curious new residential territory that was neither city nor country, but somewhere in between. Enthusiastically embraced by commentators – and much beloved in Britain ever since – the suburbs seemed to offer the best of both worlds, as Mrs J. E. Panton enthused in her 1888 book *From Kitchen to Garrett*:

> I would strongly recommend a house some little way out of London. Rents are less; smuts and blacks are conspicuous by their absence; a small garden, or even a tiny conservatory, are not an impossibility; and

if 'Edwin' has to pay for his season-ticket, that is nothing in comparison with his being able to sleep in fresh air, to have a game of tennis
in summer, or a friendly evening of music, chess or games in winter,
without expense.[71]

As Edwin's routine suggests, railways gave new definition to the distinction between work and play, one that also distanced husbands from
their wives. As thousands of Edwins or Mr Pooters left every morning
for the city, their spouses were left behind to look after the children,
choose curtain fabrics and order beef or mutton for dinner.[72] Since
middle-class 'ladies' weren't supposed to work, having at least one
domestic servant was considered essential.[73] Without even the diversion
of *oikonomia* to keep them busy, well-to-do housewives stagnated.
Industry's finest achievement, leisure, appeared as much of a curse as a
blessing.

For the vast majority of Britons, however, leisure remained a distant
dream. By 1851, 54 per cent lived in cities, and working families were
crowded into one or two squalid rooms. Most depended on casual work
– labouring for men and cleaning or sewing for women – that could be
picked up by the day. Far from being freed by railways, the poor were
effectively trapped by them: unable to afford the fares to live out in the
suburbs, where rents were cheaper but there was less work, they were
forced to live in urban slums, where they at least had access to employment, supportive neighbours and markets where they could buy cheap
food – which at the end of the nineteenth century cost them between
one half and three quarters of their income.[74] Victorian slums were the
prototypes for the shanty towns and *favelas* that surround modern cities
like Delhi, Lagos and Rio: places of transition and displacement where
rural migrants lived and dreamed of a better life.

Homes for Heroes

The efforts of social reformers to overturn the sclerotic social order were
effectively overtaken by the First World War. By decimating the male
working population, the conflict simultaneously improved women's

employment prospects and removed the domestic servant class upon which polite society had depended. At the war's outbreak, 24 per cent of women went out to work, but by 1918 the figure had risen to 37–47 per cent.[75] Crucially, the war also gave the British public a social conscience about the quality of life of the working class that had previously been conspicuous by its absence. As one member of Lloyd George's government put it, 'To let them (our heroes) come home from horrible, water-logged trenches to something little better than a pigsty here would, indeed, be criminal . . .'[76]

The result was the first and greatest house-building programme that Britain has ever seen. Overseen by the social reformer B. Seebohm Rowntree and the garden city architect Raymond Unwin, the programme transformed working-class housing in Britain, building thousands of workers' cottages (the earliest and most glamorous type of council house) with generous space standards on the outskirts of major cities. The working-class suburb had arrived – and with it the dilemmas of modern domesticity.

Unlike their Victorian counterparts, most twentieth-century housewives had no servants, yet were still expected to keep house for their husbands: in 1936, 60 per cent of British men still came home for lunch.[77] For working-class women used to fulfilling the dual role of worker and housewife, life was actually made harder by living in the suburbs, since the services they needed – jobs, markets and friendly neighbours – were no longer so close at hand. For middle-class wives, the shock was even greater: unlike their working-class counterparts, they had never been asked to cook for their husbands; indeed few had a clue what to do in the kitchen.

The post-war domestic crisis on both sides of the Atlantic was a boon for the food industry, which leaped into action, producing convenience foods that housewives could pass off as their own. US food companies led the way, making everything from Betty Crocker's cake mixes (available in a packet near you from 1940) to entire pre-cooked turkey dinners that could be reheated and served. Such subterfuges led to the creation of the adman's dream client and companion: the 'domestic goddess' who could not only turn out a roast dinner with all the trimmings, but emerge looking fabulous and fragrant in order to entertain her husband's guests.[78]

The relative lack of servants in America had in fact made the domestic role of women an issue since the early nineteenth century, leading the women's rights pioneer Catherine Beecher to demand more efficient kitchen designs and more respect for housework in her 1842 *Treatise on Domestic Economy*. 'As society gradually shakes off the remnants of barbarism,' Beecher wrote, 'a truer estimate is formed of women's duties, and of the measure of intellect requisite of the proper discharge of them.'[79]

Beecher's broadside marked the start of the American women's movement and the dawning recognition that, in a servant-less world, kitchen design mattered. Her radical designs for built-in work and storage units, published in *American Women's Home* in 1869, were the start of a revolution that culminated in Grete Lihotsky's Frankfurt Kitchen of 1926, the prototypical galley kitchen that probably influenced the one in which you made breakfast this morning, since virtually every kitchen since has been influenced by it.[80] What no amount of sleek design or nifty storage units could do, however, was square the circle of women's dual role inside and outside the home.

By 1945, over one third of British and American women were going out to work, with many doing 'men's work', as they had in the First World War, in place of servicemen fighting overseas. Despite having proved themselves perfectly capable of doing such jobs, however, women were expected to return to their domestic duties once the war was over.[81] Serene on the surface yet boiling within, the 1950s housewife epitomised the tensions of modern domesticity: the often unwilling lynchpin of the ideal nuclear family and reluctant provider of its most iconic ritual, the family meal.

The Digital Home

> A person works all day in a job he hates to buy things he doesn't need.
>
> Tyler Durdon, *Fight Club*

Few rituals express the gap between our dreams of home and the often less than perfect reality as powerfully as the family meal. The smiling parents and charming kids who shared their Sunday roasts in 1960s Bisto ads were always, to some extent, a fiction. Even in its mid-twentieth-century heyday, the family meal frequently simmered with underlying tension. Yet, however reluctant or incompetent its protagonists, there is something essential about a shared home-cooked lunch or dinner that no other meal can replicate. The preparing and sharing of food remain as fundamental to our sense of belonging as they were to our distant ancestors. For those lucky enough to have been regularly cooked for by their parents, it is family mealtimes that dominate memories of home – and our sense of having been loved.

Today, although the family meal isn't quite dead, home life in the Anglo-Saxon world is nevertheless hugely fragmented. In 2017, 31 per cent of American households with children were one-parent families, while 28 per cent of UK households were single occupancy.[82] In 2016, both partners in 61 per cent of two-parent families in the US went out to work, while in 2017 the UK figure was 72 per cent.[83] In 2010, an American gender milestone was reached, when for the first time women made up more of the workforce than men.[84] With stats like that, it's not hard to see why nobody feels they have time to cook. It's not just parents who are busy; one 2018 US poll found that teenagers were spending on average nine hours a day online, so much that even the children were worried, with 60 per cent admitting it was a 'major problem'.[85] No wonder that increasing numbers of us are resorting to buying ready meals: in Britain we love them so much that in 2017 we ate half of all those consumed in Europe.[86]

Should we fret that we no longer eat regularly together, or that our home life is disintegrating? What, indeed, is the role of home in our digital age? Is it as vital to our sense of belonging as it once was, or has its function been usurped by the 'families' we find on Facebook? Whatever the answers, there is no question that the nature of home is shifting. Once a productive hub where families lived and worked together, it is now the primary locus of individualised consumption. From the comfort of our sofas, we can shop, order food, socialise and be entertained: home, for most of us, is merely a plug-in to the global supply chains that keep the capitalist circus on the road. For many, relaxing at home is a reward for doing jobs that we'd rather not do – a deal that none of us signed up to, yet is fundamental to how modern society works.

Our farming forebears didn't want to work either, of course; work, as we have seen, was a concept that arose with the advent of agriculture.[87] The difference between most modern work and that of, say, a seventeenth-century farmer, lies in its lack of depth and variety. While in the course of a week such a countryman or -woman might have lit a fire, ploughed a field, ridden a horse, made a fence, caught a fish, woven a basket or baked some bread, most modern jobs are, by comparison, unskilled and monothematic. Typical non-professional employment these days might involve working in a call centre, stacking shelves or delivering pizza; even white-collar jobs often consist of tedious admin of the sort satirised by the BBC TV series *The Office*. Dubbed 'bullshit jobs' by American sociologist David Graeber, such occupations have proliferated in the digital age.[88] When asked by a 2015 YouGov survey whether they thought their jobs made a meaningful contribution to the world, 37 per cent of British workers said no, while 13 per cent were unsure.[89] Such work might just about pay the bills, but it can never fulfil us.

While life in a pre-industrial household was tough, it had qualities that modern life often lacks: a clear sense of purpose and belonging. Our progress towards greater material wealth has been achieved at the cost of increasing monotony and isolation, as the workplace has become progressively deskilled and dehumanised. For many, home is now a place to seek consolation in the form of companionship, entertainment and consumer goodies all accessed through a screen. Yet no matter how many friends we have on Instagram, they are no substitute for real ones in the room. As Robin Dunbar has shown, physical contact with others is essential to our wellbeing – a finding with profound implications for the way we live in the future.[90] A 2018 study by Oxford Economics, for example, found that the more we eat together, the happier we're likely to be, and that regularly eating alone is more strongly associated with unhappiness than any other factor apart from mental illness.[91]

Companionship matters to us, but as Aristotle and Maslow realised, we also need personal fulfilment. Here too, as Tibor Scitovsky observed, our consumerist culture works against us, since our pleasure in material comforts is ephemeral. Our love of gadgets is essential to the capitalist project, but is also a source of our discontent, since it denies us our most reliable joy after eating and socialising: making things.

Handy Man

Fifty years ago, the majority of jobs still involved making something, whether it was clothes, cars, ships, furniture or food. Although such jobs were industrial, many preserved some of the qualities of the pre-industrial workplace, requiring a degree of teamwork, knowledge and skill. Housework half a century ago was more craft-based too: most housewives could make pastry, bake and sew, while cars were still mechanical enough to allow those so inclined to mess about under their bonnets. Today, by contrast, few of us know how to make or mend anything. Most of us buy the things we need, and most everyday objects have built-in obsolescence. Even if we wanted to mend them, as Mathew Crawford points out in *The Case for Working with Your Hands*, many are deliberately designed to exclude us: the ultra-smooth iPhone with its tiny, star-shaped screw-heads, for example, defies an ordinary screwdriver, just as computerised 'black box' BMW engines prevent us from 'interrogating the innards'.[92]

Prompted by the incessant pressures of consumerism, most of us have got used to replacing old devices with new well before the old ones stop working. Yet our throwaway culture damages more than our planet; it threatens something essentially human. As Crawford notes, the kind of creativity required to mend things involves a highly sophisticated cognitive effort that brings its own special reward. The fact that our brains are wired to get pleasure out of such manual tasks is hardly surprising; we have, after all, been co-evolving with tools for some 3.5 million years. Our ancestors puzzled things out, not just with their brains but with their hands, a fact reflected when we speak of 'reaching' for an answer or 'grasping' an idea.[93] Working with our hands is as innate to us as thinking with our brains, two functions which, as Crawford points out, form two indivisible halves of our material consciousness.

Making things gives us joy because it brings body and mind together – it engages the whole of us in a way that few activities do. It is satisfying because it is useful. It locates us in the world, giving us something that we can point to and say, 'I did that.' One can only imagine the sense of belonging that such a feeling must have given medieval craftsmen, living somewhere that they had had a big part in making. Modern farmers feel the same about their land, surveying territory shaped by past generations of their family. Our very idea of home grew out of such craft and graft: the sense of connection to a place that one has helped to create. Few of us today have the chance to build our own homes, yet most of us probably remember making something at school – a wonky pot or an advent calendar out of fuzzy felt – and can vouch for the satisfaction of bringing it home to show our parents and the inner glow of having made something useful, beautiful and good.

It is rare to experience such joy in our day jobs nowadays; creativity and craftsmanship have largely been edited out of the workplace. But what of home? Surely we have the power to spend our leisure time as we please? For the Greeks, flourishing was all about active engagement with the world, not escaping from it. Sport mattered because it prepared one for action and sharpened the mind; to distract the mind while ignoring the body – as we do when online – would have been unthinkable to them. Today, with robots doing much of our work and algorithms anticipating our every desire, we have little physical reality left with which to interact. It's not hard to see why the virtual world can seem more real to today's children than the real one: their experience of it is far more frequent, tactile and immediate. The result is not just an ignorance of nature and the loss of social skills, but an alarming physical decline: recent studies have revealed children to be more short-sighted and to have a weaker sense of balance than their pre-digital counterparts.[94]

Post-industrial society is the very opposite of the wilderness where we evolved. Instead of providing the friction that stimulated our ancestors' engagement and ingenuity, it stalks us with cookies and loyalty cards that anticipate our every need. Learning to survive in the wild was how our species became human; now we live in a digital hall of mirrors that merely fuels our narcissism. Can we flourish in such a world, or is it time to rethink our idea of what a good life is?

Paradise Lost?

The Greeks believed that a good life required struggle; without it, they thought, being human had little meaning. They admired hard work and frugality because they were necessary to becoming a good citizen. From modern political rhetoric in the UK and US, you would think that we thought the same – 'hard-working families' are constantly cited as the ideal – yet it is wealth that we dream of these days, not *oikonomia*. This was probably also true in ancient Athens – Aristotle wouldn't have mentioned *chrematistike* if he didn't think it was a threat. It's only human to want an easy life. Yet, as our leisure choices betray, we contemporary urbanites yearn for some sort of action or challenge. We regularly go on adventure holidays or abandon the comfort of our homes to camp under the stars, light fires, catch fish or just barbecue sausages in the rain in order to remind ourselves what it means to be alive.

Although we can never experience life as a hunter-gatherer, we have the same bodies, minds, physiological and psychological needs as our ancestors – as well as the same dependency on nature. Before they vanish from earth completely, therefore, is there anything we can learn from our forager cousins about how to be at home in the world? Certainly Colin Turnbull's account of the Mbuti suggests a people living in remarkable harmony with their environment, a way of life that is now under severe threat from overpopulation, logging, civil war and soaring demand for bush meat in their native Congo.

If we see ourselves as somehow diminished in comparison to people like the Mbuti, we wouldn't be the first. The term noble savage, coined by John Dryden in 1672, sums up the mixture of awe and admiration (with a tendency to patronise) that those living in cities have felt for those living in the wild at least since Roman times.[95] For as long as humans have farmed, we've greeted each new technological advance as both a blessing and a curse: a blessing because it has relieved us of hard work, and a curse for the very same reason.

Our greatest sacrifice in our long transition towards urban civilisation has arguably been a loss of contact with the world – and thus of agency. By outsourcing whatever it takes to keep us alive, we have swapped one set of skills for another: the ability to read the landscape, make arrows and track game, say, for those of programming computers, texting and searching the Internet. Although both sets of skills are invaluable in their context, they are fundamentally different in nature: while the former are directly linked to survival, the latter are at several removes. A hunter-gatherer is directly responsible for his or her life, while we must rely on numerous strangers to survive. Despite technological capacities that would have made our forager ancestors gape in wonder, therefore, keeping our own lives on track is utterly beyond our grasp, as we discover to our frustration whenever our computers say no.

The divisions of labour and knowledge that have given us urban civilisation have made us far greater than the sum of our parts, but, like so many aspects of progress, they have come at a cost. Compared with the average hunter-gatherer, we are less in tune with our surroundings and observant of change, less able to feed and protect ourselves and generally far less self-reliant. Although we have plenty of secondary stimulation in our lives, we are starved of primary engagement with the world.

The Virtue of Necessity

> Necessity is the mother of invention.
> English proverb

One thing that still links us directly to the world is of course food. Our evolution has been driven by it. As we have found new ways of feeding ourselves over millennia, the places we call home have changed beyond recognition; yet, to a remarkable extent, the stuff we eat has remained the same. Despite newcomers to the gastronomic scene like Soylent and Krispy Kreme, we mostly subsist on the same animals, grains, pulses, tubers, nuts and greens that our ancestors ate – albeit modified to within an inch of their lives.

It is no accident that our digital era has seen a parallel upsurge of interest in food-related crafts in industrial nations like the UK and US. Whether it's butchery classes, pickling, brewing or making sourdough bread, the desire to get down and dirty with food has never been stronger. Waiting lists for allotments in the UK are longer than at any time since the Second World War, and sales of seeds for fruit and vegetables have outstripped those for flowers. Independent bakers, cheese makers, brewers and coffee houses have also made a comeback on both sides of the Atlantic, to some extent reviving the working households that industrialism swept away.

Although this foodie resurgence is partly in response to the blandness and destructiveness of industrial food, it is also symptomatic of something far deeper. Working with food is the perfect antidote to living in our virtual, dematerialised world. Since it consists of living things that come from nature that we need in order to survive, food reminds us of the very thing our modern lives are designed to obscure: the virtue of necessity. Food is something we can make, that brings us together and grounds us. Growing, cooking and preserving food are all manual skills we can get really good at, earning us plenty of friends in the process. Food, in short, is something through which we can root ourselves in the world, both socially and physically.

Could we take this resurgence of interest in food to rethink our idea of home? Some have already done it. BedZED, a pioneering mixed-use sustainable development completed in 2002 in the London Borough of Sutton, included generous gardens, balconies and allotments as part of the original masterplan, in order both to minimise residents' carbon footprints (BedZED stands for Beddington Zero Carbon Energy Development) and to help foster a sense of community.[96] Both strategies seem to have worked: by 2009, a study found that the average BedZED resident had an ecological footprint 19 per cent lower than those living in surrounding neighbourhoods and was friendly with twenty of their neighbours, compared to just eight for those living nearby.[97] When asked what they most liked about living in BedZED, residents said it was the sense of friendliness – it felt, they said, like living in a village.

With its high-density, low-impact, mixed-use approach, BedZED suggests how we might all thrive in a zero-carbon economy. Bioregional, the charity that developed the project in collaboration with the architect Bill Dunster, calls it One Planet Living. The idea is that, by combining high-quality eco-design with generous public and private spaces to live, grow, work and play, we can recapture the sense of community that once animated rural villages and urban neighbourhoods, creating a way of life that people actually want to live. With its productivity, self-reliance and collaborative spirit, BedZED reinvents many of the most positive aspects of the working household. With all residents able to work at or close to home, the question of who cooks dinner or looks after the kids becomes less fraught, while regular veg-box deliveries take some of the heat out of shopping. Such village-like living also demonstrates how rethinking our idea of home can help us ditch our consumerist habits, by offering us something far better: an active, pro-ductive, sociable, natural life. Expanded to the scale of the city, such thinking could be transformative. As the Greeks understood, happiness and resilience start at home.

Whatever our homes look like in the future, food will be at their core. For social animals like us, the sharing of food will always be central to bonding with others – and thus to our sense of feeling at home. From our very first meal to our last, food and love are inextricably fused in our brains. Throughout our lives, the chance to show our love through food lies in every meal that we grow, cook and eat. It is the basis not just of a good life, but of being human.

4
Society

375 — *Paris.* - Les Halles.
Arrivée des Marchandises.

A Tale of Two Markets

> How can you govern a country which has two hundred and
> forty-six varieties of cheese?
>
> Charles de Gaulle[1]

It is five in the morning and barely light, and I am driving through a
vast industrial compound south of Paris. Massive sheds rear out of the
gloom, their tarmac forecourts crowded with articulated lorries either
waiting to offload, backing into loading bays, lights flashing and klaxons
bleeping, or queuing to join the stream of traffic heading for the exits.
The scale and sense of urgency remind me of an airport, yet what is
being shifted here isn't passengers, but food: 24,000 tonnes of it per day,
to be precise. Welcome to Rungis, the world's largest wholesale fresh-
food market.

Everything about Rungis is big. Its 234-hectare site is comfortably
larger than the Principality of Monaco. In 2018, it employed 12,000
people and supported a further 102,000 jobs throughout France, generating an annual turnover of €9 billion – 0.33 per cent of France's total
GDP.[2] Its eight fruit and vegetable halls provide a total of 3.7 kilometres
of linear selling space, shifting 1.2 million tonnes of fruit and veg a year
– around half of Parisians' five a day. Delivering all this produce generates 1.5 million vehicle journeys every week. During the morning rush
hour, hundreds of porters, barrow boys and forklift drivers weave their
way between stacks of pallets in a frenetic, scaled-up version of the
opening scene in *Oliver!*

Less hectic but no less impressive are the dairy halls which, as one
would hope and expect, house the world's biggest cheese market.
Rungis sells 65,000 tonnes of *fromage* a year, from hulking, 100-pound

rounds of Comté to delicate, ash-dusted, volcano-shaped goats' cheeses carefully packed in straw-filled crates. When I ask the director how many varieties of cheese he sells altogether, he simply smiles and gives me that special French shrug that means 'Who knows?' '*Milles!*' he says pleasantly, mimicking the myriad sizes and shapes of lactic provender with his hands.

While hard cheeses can lurk in Rungis' maturing rooms for many months, operations down the road at the market's 24,000-square-metre fish hall, the *Marée*, are all about speed. Each night, 1,500 tonnes of salmon, sole, lobsters, oysters and sea-urchins arrive and depart in a matter of hours. Packed in ice and sped through the market by white-coated porters as though in some medical emergency, the fish briefly transform the hall into a glistening mirage of the sea that by dawn has already vanished, leaving just a trace of its presence in the mineral under-the-pier smell that drifts in the freezing air.

Much earthier are the meat halls, where the red, white and purple carcasses of cows, pigs and sheep hang suspended from acres of metal rails. Some 270,000 tonnes of gigot, fillet and entrecôte were sold here in 2017, worth €1.5 billion between them. This being France, there is even a hall dedicated to tripe, of which more than 20,000 tonnes were sold that same year. Despite the onward march of McDo, some aspects of French gastronomy remain intact, at least for now.

Rungis is a leading example of an institution increasingly rare in the West: a market that plays a key part in feeding a major city. Although it has international reach, most of the food here is destined for the markets, shops and restaurants of Paris and the surrounding region. The market may not be located in the city centre, but it nevertheless represents the city's guts. The powerful bond between markets and cities is obvious if you stroll through any town built before the railways: you are sure to find a marketplace at its core. Today, most are far more likely to be filled with tourists than traders, yet such places remind us of the profound role played by food in shaping urban life.[3]

From the Athenian agora and Roman forum to London's Smithfield and Amsterdam's Dam Square, markets have shaped history. Democracies have been born in them, empires built, revolutions ignited and kings crowned or dethroned. For centuries, markets have been the backdrop against which all of public life has played out. Yet their power has ultimately derived, not from the pomp and circumstance of state, but from their everyday role as places where people came to buy food and swap news. Above all, markets were spaces of sharing and encounter, where the inner workings of the city became visible. The town hall on the market square overlooked by a nearby church or mosque is an urban archetype repeated all over the world: commerce under civic control, beneath the watchful eye of God. More than any other space, markets have been where the power lines of society have conjoined.

Fifty years ago, Paris's food market was still operating in the heart of the city, in the famous glass and iron halls of Les Halles.[4] Dubbed *Le Ventre de Paris* (the Belly of Paris) by Émile Zola, the market not only supplied all the fruit, vegetables, butter, cheese, fish, shellfish, meat and game that Parisians could eat, but all the news, gossip and wit they could digest too, as Zola's description of an early morning there suggests:

> The piles of vegetables on the footpath began to overflow onto the road. The traders had left narrow gaps between the various piles to let people pass. The wide footpath was covered from end to end with dark mounds. So far nothing could be seen as the lanterns swung by, except the luxuriant fullness of the bundles of artichokes, the delicate green of the lettuces, the coral pink of the carrots, and the smooth ivory of the turnips ... The footpath was now full of people: a whole crowd was moving around among the goods on show, chattering and shouting. A loud voice in the distance cried 'Endives! Endives!' The coming and going between the roadway and the market grew more intense as people collided and cursed, while all around there was a clamour of voices growing hoarse because of prolonged wrangling over a sou or two. Florent was surprised by the calmness of the farm women with their bright headscarves and tanned faces amidst all this jabbering.[5]

Established in the twelfth century conveniently close to the Grève, the city's main grain port on the Seine, and the fortress at the Louvre castle, Les Halles was soon the bustling centre of the Paris food trade.

By the eighteenth century, it had grown into a fortress-like stronghold with its own rules, clocks and mafia-like trading dynasties. Unlike in London, where markets were spread all over the city, the Paris food trade was concentrated in one place and controlled, at least in theory, by a 'grain police' which answered to the king. The reason for this was partly geographical: while the Thames was navigable by ocean-going ships, allowing London to import its food from wherever it pleased, the Seine was not, with the result that the capital was forced to get its food, at musket-point if necessary, from the surrounding countryside.[6] The dangers of this set-up were exposed during the 1780s, when a series of failed harvests led to food shortages and riots, the latter incited partly by Les Halles porters who blamed the king for the famine.[7]

As Louis XVI discovered to his cost, being held responsible for feeding people, especially when supplies are unreliable, is a perilous duty. Small wonder that leaders all over the world have shunned the responsibility wherever possible – English monarchs, it should be noted, were happy to let London feed itself. In eighteenth-century Paris, the clash between food and politics sparked revolution, yet just half a century later, the arrival of the railways would change the relationship for ever.

By making it possible to transport food rapidly over great distances, railways obliterated the geographical constraints that had hitherto hampered the growth of cities. They began to expand with immediate effect, and as they did, the ancient role of markets started to fragment. No longer would cartloads of cabbages be trundled to market by gnarled countrywomen in headscarves; from now on, cities would be fed in an entirely new way. Markets would be replaced by shops supplied by producers and distributers whose focus would be increasingly global. Like many great markets, Les Halles lived on well beyond its sell-by date, eventually succumbing to the wrecker's ball in 1971, the same year that London's Covent Garden narrowly escaped a similar fate. Today, the old market site is occupied by the Forum des Halles, a soulless sunken shopping mall whose swooping beige roof does little to ameliorate its baleful effect on this once-beating heart of the city. With Paris's brassicas and Brie redirected to Rungis, the essential bond between city and country that is the mainspring of civilisation became invisible and increasingly unaccountable.[8]

The Virtual Market

Today, most of the trade in Paris takes place not in the city but in the steel-and-glass financial archipelago at La Défense – which is to say, it happens in cyberspace. The transformation has been rapid: just a generation ago, futures traders in striped blazers still jostled and bellowed at one another across the sweaty 'pits' of trading floors, risking life, limb and larynx to secure a deal. Nowadays, however, the biggest deals, whether in finance, food or other commodities, are made in air-conditioned bunkers by mega-computers that merely hum as they process the trillions of algorithmic calculations that can earn or lose millions of dollars in the blink of an eye.

The move from human to digital high-frequency trading (HFT) is merely the latest in a series of shifts in ways of buying and selling that stretch back over two centuries. In Paris, the trail begins in the rotund form of the Bourse de Commerce, built next to Les Halles in 1763 as a corn exchange to replace the muddy market at the Grève. Its state-of-the-art design sent England's leading agronomist, Arthur Young, into raptures:

> It is a vast rotunda, the roof entirely of wood . . . it is as light as if suspended by the fairies. In the grand area, wheat, pease, beans, lentils are stored and sold. In the surrounding divisions, flour on wooden stands. You pass by staircases doubly winding within each other to spacious apartments for rye, barley, oats etc. The whole is so well planned and admirably executed, that I know of no public building that exceeds it in France or England.[9]

The Bourse traded for just over a century before being converted into the Commodities Exchange in 1885. From now on, rather than being stored and sold on site, food would be traded remotely through new types of contracts known as futures. Pioneered by the Chicago Board of Trade in 1864, these contracts were designed to buffer farmers and buyers from extreme price fluctuations in the newly railroaded Midwest, where grain gluts from bumper harvests, interspersed with demand surges from the Civil War, were causing market havoc.[10] By agreeing a price ahead of the harvest, everyone could hedge their bets, assuring farmers of a fair deal for their crop, while allowing buyers to speculate on the market.

Futures contracts now dominate, not just trade in food, but in pretty much everything. Known as derivatives because they're derived from an underlying asset, they established the principle that, instead of buying a commodity for its use, one could simply gamble on its future price, thus creating a meta-market based on pure speculation. Today, around 80 per cent of traders who deal in food futures have no intention of either producing food or taking delivery of it.[11] The proliferation of such non-food investors dates back to the deregulation of US commodities trading in 2000, when 'over the counter' (OTC) derivatives (deals made directly between two parties rather than through the exchange) were exempted from scrutiny.[12] The result was an explosion of speculative trading and interest in commodity index funds (CIFs), unregulated investments that follow an index of selected commodities, such as grain or livestock. From 2003 to 2008, CIF holdings mushroomed from $13 billion to $317 billion, helping to fuel the market volatility that preceded the 2008 crash. Wheat and rice prices over the period increased by 127 and 170 per cent respectively, while maize prices nearly tripled, leading to food riots in some 30 countries.[13] Far from stabilising food prices, futures markets were now adding to their volatility.

As Oliver Raevel, former head of commodities at Euronext, Europe's largest commodities exchange, explained when I went to meet him in 2016, managing such volatility was his primary mission. Sitting in his office high above the central plaza at La Défense, he showed me the latest graphs from the United Nations Food and Agriculture Organization, projecting the global per capita decline of arable land over the coming decades. 'The world faces a double challenge over the next thirty to forty years,' he told me, 'since the additional two billion mouths to feed will all be urban.' As a result, said Raevel, crop yields would need to increase by 80 per cent; his job, as he saw it, was to help farmers achieve that goal.

Food is unique among commodities, Raevel explained, since the usual rules of supply and demand don't apply. Due to the vagaries of weather, harvests vary widely from year to year, while demand stays comparatively constant. Left purely to the market, therefore, prices fluctuate wildly, with farmers perversely doing *better* in years when crops fail and food shortages push prices up. On top of this, farmers have to make big decisions each year about which crops and how much to sow. If the market is volatile, they may not risk growing as much as they might if prices were stable, thus reducing the overall amount of food produced. Futures contracts help with all this, said Raevel, because they give farmers 'price discovery'. Yet such contracts are also somewhat old-fashioned, since their underlying asset is physically deliverable. Unlike gold, for example, which can be bought and sold at any time, grain is a living substance that must be harvested when it is ripe and that someone will eventually eat.[14]

This link to physical reality means that food futures have to be constantly tweaked in order to reflect real-life conditions. If there are heavy rains at harvest time, for example, the value of a crop may be lower than specified in the contract, due to its lower protein yield. In order to minimise each party's risk, therefore, contracts are constantly 'marginalised', meaning that each party's margin or deposit (up to 20 per cent of the contract value) is adjusted to reflect current market conditions. In addition, the parties can agree at any time to end their contract through an 'exchange for physical' (EFP) deal, in which the stock is sold at its actual 'spot' market rate, allowing the buyer to secure his supply, perhaps for a higher price than in the original contract but with the guarantee of delivery. Such deals are fundamental to the way the exchange operates: just 1 per cent of futures contracts are actually held up to the date of expiry.

Listening to Raevel talk about grain, rain, sun and soil up in his glassy eyrie is something of an out-of-body experience. It all feels a long way from the reality of food: apart from a few stalks of wheat in a pot on his desk, we may as well be talking about car insurance. Only at the top of the building does one get a glimpse of what Euronext is really about: a fully glazed, double-height space in which twenty or so people sit staring at monitors beneath a huge, Piccadilly-Circus-like circular screen displaying the latest market prices around the world. This is where the exchange does its work, monitoring the market and relaying up-to-the-minute data to its many clients. Compared to the magnificence of the

Bourse, the space is underwhelming, yet the volume of trade passing through here is millions of times greater than anything the old exchange ever handled.

Every year, some $1.5 trillion worth of food – one quarter of the world's supply – is traded on such markets.[15] Should we worry that our most vital commodity is distributed via what amounts to one vast, globalised casino? The human rights lawyer Olivier De Schutter, who served as the United Nations' 'special rapporteur on the right to food' (2008–14), certainly thinks so. In a 2010 briefing note, De Schutter blamed speculation as a key factor in the 2008 food price spikes, concluding that 'fundamental reform of the broader global financial sector is urgently required in order to avert another food price crisis'.[16] In a world in which financialisation is increasingly seen as a panacea, however, such reform in the near future seems unlikely.

Since the millennium, deregulation and automation have transformed the way that the global economy operates. With the advance of digital finance, old distinctions between commerce, government and civil society have become blurred. Wealth and power, once manifest in the physical fabric of the city, are now hidden from view. Like the virtual exchange at Euronext, the true reach of institutions and corporations is intangible and virtually unlimited. In contrast to the historic marketplace, where traders sold their wares under the watchful eye of civic authorities, our modes of exchange in the digital age are fleeting, secretive and obscure.

What implications do such economic structures have for our chances of freedom, opportunity and justice? How, in the modern world, can we perceive the power structures that govern us, let alone challenge them? The spatial transformations wrought by industrialisation have been augmented by a digital disembodiment that renders power and influence all but invisible. So rapid and radical has this transformation been that we are only just starting to grasp its implications.

The very nature of the public realm is shifting, and with it the exchange of ideas and goods at the heart of society. Once a physical place where anyone was free to act, public space was essential to the evolution of democracy. From the Athenian agora, where citizens gathered to vote and debate, to seventeenth-century London coffee houses where public opinion and the modern press evolved, such spaces have hosted the discourse that has shaped free society.[17] Now, however, the space of exchange is increasingly moving online, to digital platforms which, as the illegal targeting of Facebook users during the 2016 Brexit vote demonstrated, is far from public or democratic.[18] If we are to thrive in this new reality, we are going to need ways of making our relationships tangible again, which means examining what it means to share.

A Good Society

Humans, as Aristotle noted, are political beasts. Living in groups comes naturally to us; we belong in society. By working together, we have become more than the sum of our parts, yet cooperation has also required us to evolve strategies for sharing, a process that gave us, among other things, language and economy.

Sharing is fundamental to every society, yet what, precisely, do we share? At the scale of the planet, we clearly share natural resources, not just with our fellow humans, but with all other species – a condition that, as the authors of Genesis were at pains to point out, brings its own responsibilities. Closer to home, we share territory, ideas, values, language, knowledge and skills: the building blocks of our local culture. In our inner circle, with any luck, we also share love, a virtue that, as the Christian command to 'love thy neighbour as thyself' suggests, is key to how we *ought* to share all the rest.

The most important material resource that we all share is, of course, food – along with the land, sea, water and other resources needed to produce it. How to share food equitably has always been a question fundamental to that of how to make a good society. For early humans, the question was relatively simple. Finding food, keeping warm and fending off predators would have been high on the agenda – not much room for ambiguity there. Sharing fairly would have been integral to the group's survival, the essence of their sense of cohesion and trust. In more evolved societies, however, the question became more complex. Once it was possible to fall out, or even kill, over ideas as well as portion sizes, human society could be said to have come of age.

As Yuval Noah Harari has argued, ideas matter to us. Our ability to tell stories is vital to our capacity to cooperate, since it is through shared myths that we gain our sense of commonality. As Harari wrote in *Sapiens*, 'Any large-scale human cooperation, whether a modern state, a medieval church, an ancient city or an archaic tribe – is rooted in common myths that exist only in people's collective imagination.'[19] So essential are such shared fictions that we barely notice them, yet they are the basis upon which, for example, lawyers contest a court case or accountants work on the same balance sheet. Even when we agree to meet friends for dinner at 7 p.m., we are making use of the convention that days are divided into twenty-four equal bite-size chunks called hours.

Since such ideas are vital to society, one might conclude that an ideal world would be one in which we all thought the same. Yet a moment's reflection reveals how dreadful it would be if we all wanted the same jobs, houses, holidays and partners – the stuff of a dystopian nightmare. Diversity is the essence of a good society, yet this poses another dilemma: if we all pull in different directions, how can we find consensus? In order for society to function at all, there must be some sort of common goal towards which everyone can theoretically strive. So what might that be? What do we all *want*?

For Aristotle, the answer was of course happiness, yet, as he himself admitted, there are likely to be as many views of what that means as there are people. Nevertheless, he insisted that creating conditions under which everyone could flourish was the ultimate goal of politics. Society is born of compromise, its rules and boundaries shaped by countless negotiations that seek to establish common good. At home we make up such rules all the time: I'll sleep on the right, you sleep on the left, don't nick the duvet. In public, however, the rules by which we live are mostly inherited. What, when and how to eat, how to dress, behave and speak, what is good or bad, legal or forbidden, public or private, polite or rude, when to attend school, how to worship, who gets to vote, which side of the road to drive on: all are passed down by the societies into which we are born. When we grow up, we may challenge such rules, but by then they will already have shaped us. Society, like geography, is embedded in us.

As Socrates discovered the hard way, there is a balance to be struck between accepting established norms and endlessly challenging them. He was the first to acknowledge that living in society means abiding by its rules, even if they are not to one's liking. Jean-Jacques Rousseau agreed, pointing out in his 1762 book *The Social Contract* that freedom comes at a price. In order to enter society, said Rousseau, we must give up our individual rights, yet since everyone else must do the same, we regain our freedom as citizens: 'each giving himself to all, gives himself to no one'.[20] Since the system only works if everyone submits to it, Rousseau came to the somewhat startling conclusion that 'whoever refuses to obey the general will shall be constrained to do so by the whole body; which means nothing else than that he shall be *forced to be free*' (my italics).[21]

Individual freedom and social cohesion, as Rousseau realised, always exist in tension. So where does that leave us as would-be good political animals? Selfishness and altruism, as we saw earlier, are both innate in us, dating back to the subtle manoeuvrings of tribal life. Could learning to balance the two therefore be the key to happiness? The American psychologist Shalom H. Schwarz believes so. In a 2006 study conducted across sixty-seven countries, Schwarz and his team asked people what they most valued in life, encompassing a wide range of attributes including power, security, tradition, achievement, stimulation and so on. He found a remarkable consistency: people's values were generally centred within two sets of extremes, between selfishness and altruism on the one hand, and novelty and conservatism on the other.[22]

As Tim Jackson argues in *Prosperity Without Growth*, Schwarz's findings show why the pursuit of economic growth (*chrematistike*) can never make us happy. In a modern democracy, says Jackson, the state tries to ensure stability and prosperity by managing the economy, which it does by balancing precisely the tensions identified by Schwarz. But when the government's goal is recast as pure economic growth, selfishness and novelty will always triumph over tradition and altruism, putting society in a state of self-defeating conflict. In order to thrive, Jackson suggests, we need to bring altruism and stability back into economics. Only then will our economy have a hope of delivering what, deep down, most of us really want: to live in a society that balances innovation with tradition and which is both prosperous and fair.

Seventy thousand years since humans spread out from Africa, our dilemmas have barely changed. We've created a remarkable diversity of social forms between us: democracies and dictatorships, secular states and caliphates, with many variants in between. Such societies have coexisted for millennia, sometimes in isolation, often trading, frequently at war. In our increasingly interconnected world, however, our coexistence has taken on new significance, as global threats require us to cooperate as never before. The task is daunting, yet there are clues in our past that might help us, since at one time we were all natural collaborators, when we lived around the fire.

You Hunt, I Cook

> Government, like dress, is the badge of lost innocence.
>
> Thomas Paine[23]

Sometime in the distant past, our ancestors grasped that by working together they could all lead better lives. The invention of language was key to this, as was the division of labour. Men going off to hunt and women staying behind to cook was, as we have seen, the prototypical social contract, a leap into the abstract that no other animal has consciously replicated. Such contracts formed the basis of early societies, raising the question of how to share the fruits of everyone's labour. The problem goes to the heart of politics, since it raises the issues of fairness and value. How does one compare, say, tasks such as picking berries, fetching water or cooking a meal with those of catching a fish or spending all day chasing an antelope that may ultimately elude you?

As anyone who's pitched a tent in company will no doubt recognise, dividing such labours fairly can make the difference between a happy and an unhappy camp. Yet assigning value to different tasks requires a high degree of sophistication, especially when they involve varying degrees of effort, skill or risk. For our ancestors, such judgements would have been reflected in the way in which the reward for such efforts – the evening meal – was shared. Among the contemporary Hadza of Tanzania, for example, foods that are particularly delicious, nutritious or hazardous to acquire such as meat and honey are highly prized, their value reflected in the way they are ritually eaten. Successful hunters celebrate their kill by cooking the animal in front of the whole group, then carving it up and presenting the biggest share to their family before passing the rest around.[24] The Hadza may not pay for their food, yet its inherent value is obvious to everyone.

As well as being our oldest form of economy, the shared meal is arguably the most sophisticated. Direct, transparent and adaptable, its dividends rely not on prices or markets but on instinctively held common values. Which is ideal if one happens to live in a group of twenty or so. But what happens when society scales up? Hunter-gatherer bands essentially function through consensus. Although there may be a 'big man' who leads through sheer force of personality, there is no formal leader. As Robin Dunbar has pointed out, such an arrangement demands a high degree of social intelligence, since it requires each member of the group to maintain personal relationships with everyone else. There is a natural limit to the number of such relations that most people can manage, Dunbar found, the average being around 150, a figure now known as 'Dunbar's number'.[25]

Human social capacity thus limits the size to which informal societies can grow. Beyond this point, bands tend to appoint a formal chief, with a suitable bureaucracy to support him or, very occasionally, her. The band thus becomes a chiefdom or tribe, at which point everything changes. Consensus is replaced by authority; rules are imposed from above; and a sense of common identity is reinforced, not through shared meals, but through laws, slogans and symbols. Evidence of such chiefdoms can be found from around 5500 BCE, a time when advances in agriculture made it easier to feed an unproductive ruling class.[26] Their emergence thus coincided with the rise of settled farming communities, whose new social divisions – between leaders and the led and feeders and the fed – paved the way for the organisational hierarchies that would characterise cities.

Our social evolution suggests that increasing inequality is the price we must pay for civilisation, but is that necessarily so? Certainly it is hard to ignore the fact that all great civilisations, from ancient Greece and Rome, medieval China and the Ottoman empire to Britain and the United States, have been based on slavery. In the world's first democracy, Athens, one in three people was a slave.[27] Plato considered this entirely natural, since his ideal republic was ruled by a philosopher class, who would clearly need plenty of slaves to do their farming, cooking and washing. Aristotle appears rather less convinced, asserting in an uncharacteristically muddled passage in his *Politics* that some people were born to be ruled and were thus 'slaves by nature'.[28]

The fact that thinkers as great as Plato and Aristotle could accept slavery as 'natural' shows how powerfully our ideas can be moulded by culture. Both philosophers were proud citizens of Athens who saw others as less deserving than them. In fact, most Athenian slaves *were* others, in the sense that they were foreigners captured in war. Today, with global migration on the rise, the tensions latent in national pride and prejudice are being exposed. Populism and nationalism are gaining ground, as people react to the failed promise of capitalism and blame migrants for their fate. The irony is that low-paid migrant workers are themselves the product of capitalism, fulfilling the roles once performed by slaves. If we are to build a society fit for the twenty-first century – which is to say one based on collaboration rather than exploitation – we're going to need a better mechanism for sharing. To explore what that might be, we first need to understand how we ended up with the one we've got, which means studying the political idea that it evolved to serve, democracy.

State of Nature

> Democracy is the worst form of government except for all those other forms that have been tried from time to time.
>
> Winston Churchill[29]

Democracy is arguably Western civilisation's crowning achievement. The word is derived from the Greek *demokratia* (*demos*, people + *kratia*, rule), a departure from *aristokratia*, or rule of the elite, which had previously prevailed. Although Athens declared itself a democracy sometime in the sixth century BCE, we would hardly recognise it as such today; it was a patriarchal society in which slaves and women were denied any public role. It is nevertheless some measure of the Athenian achievement that two millennia would pass before anyone would seriously attempt to examine the principles of democracy once again.

That man was the Dutch jurist Hugo Grotius, who in his 1625 *De Jure Belli ac Pacis* (*On the Law of War and Peace*) argued that all humans were born free and equal, with the natural right to defend themselves against others. Grotius' concept of natural rights may seem obvious now – at least to anyone living in a modern democracy – yet in his day it was considered highly incendiary, not least because it challenged the divine right of kings. By suggesting that people sacrificed individual sovereignty in order to enter society (an idea later taken up, as we have seen, by Rousseau), Grotius proposed that ultimate power resided not with the monarch or even with God but with the people themselves. Apart from forcing him into exile, this idea posed a new conundrum: if power resided with the people, how should they come together to form society?

Among the first to attempt an answer was Thomas Hobbes. Having been abandoned by his alcoholic clergyman father as a boy and having lived through the English Civil War, Hobbes's view of his fellow humans was understandably somewhat jaundiced. The opening passages of his 1651 book *Leviathan* amount to a pitiless critique of his own species. 'Such is the nature of men,' he wrote, 'that howsoever they may acknowledge many others to be more witty, or more eloquent, or more learned; yet they will hardly believe there be many so wise as themselves.'[30] Humans were *capable* of reason, Hobbes continued, but their judgement was clouded by their susceptibility to violent passions, which led them astray. They were prone to mistake fantasy for truth, not least because they rarely bothered to think things through from first principles, but rather relied on received wisdom from others. 'He that takes up conclusions on the trust of Authors,' warned Hobbes, '. . . loses his labour, and does not know anything, but only beleeveth.'[31] Humans, in short, were emotional and credulous, and thus prey to various forms of ABSURDITY (spelled in capitals for emphasis).

When Hobbes went on to imagine this sorry lot in what he called a 'state of nature', the result was, as you might expect, far from pretty.[32] People would have been at each other's throats, said Hobbes, since everyone would want the same things – power, wealth and glory – and since they couldn't all have everything, they would 'endeavour to destroy, or subdue one another'.[33] The result would be perpetual war, in which people would live under 'continuall feare and danger of violent death'.[34] Life, said Hobbes, in his famously dismal summary, would be 'solitary, poor, nasty, brutish and short'.

The only way that such creatures could cohabit peacefully, Hobbes concluded, was to create an absolute sovereign – the eponymous Leviathan – with absolute authority to subdue its subjects, if necessary by the sword. People would gladly submit to such a regime, he argued, since the alternative would be so much worse.[35] *Leviathan*'s dramatic frontispiece, in which a vast crowned figure bearing a sword and crosier looms over a miniature domestic landscape, encapsulates Hobbes's idea. At first sight, the figure seems to be wearing chainmail, but on closer inspection he turns out to be made up of thousands of tiny bodies: the Leviathan, in other words, is a literal 'body politic', whose power resides in the people yet whose authority makes it akin to a 'Mortal God'.[36]

Not least due to this blasphemy, *Leviathan* was met with outrage similar to that which had greeted Grotius' thesis not long before, and its author only narrowly escaped being tried for heresy. Far less controversial was the work that was in many ways its natural counterpart, John Locke's 1690 *Second Treatise of Government*. Written forty years after *Leviathan* in the peaceful aftermath of the Glorious Revolution, Locke's work was rather more optimistic than its predecessor.[37] Indeed, when Locke followed Hobbes's lead by imagining humans in a state of nature, the result could hardly have been more different:

> The *State of Nature* has a Law of Nature to govern it, which obliges everyone: And Reason, which is that Law, teaches all Mankind, who will but consult it, that being all equal and independent, no one ought to harm another in his Life, Health, Liberty or Possessions.[38]

Like Aristotle before him, Locke believed that men were possessed of reason, and were therefore perfectly capable of governing themselves. Their main task, it seemed to him, was to work out how to share the earthly bounty that God had clearly intended for everyone: 'The Earth, and all that is therein, is given to Man for the Support and Comfort of their being. And though all the Fruits it naturally produces, and Beasts it feeds, belong to Mankind in common . . . no body has originally a private Dominion, exclusive to the rest of Mankind, in any of them.'[39]

Earth belonged to everyone, said Locke, which raised the question of what gave anyone the right to call something his own. The answer, he reasoned, was that since God obviously didn't intend anyone to starve, people must have the right to feed themselves, and thus to possess the food they needed, such as by taking it directly from the wild: 'The Fruit, or Venison, which nourishes the wild *Indian*, who knows no Inclosure, and is still a Tenant in common, must be his, i.e. part of him, that another can no longer have any right to it.'[40]

The need to eat – as commanded by God – was thus the means by which people earned the right to possession. Food became theirs the moment they removed it from a state of nature, whether by plucking fruit from a tree or by felling a deer. Food belongs to us individually, said Locke, the moment we apply labour towards acquiring it. Furthermore, it is our moral duty to apply such labour to the land, since by doing so we increase its yield, so there is more to go round. God didn't put us on earth just to laze about, said Locke; on the contrary, he made it for 'the use of the Industrious and the Rational', so that 'it cannot be supposed that he meant it should always remain common and uncultivated'.[41]

By cultivating the earth, Locke reasoned, we effectively make a gift towards our fellow men, since we reduce the amount of land needed to sustain us, making more available for others. The only caveat – albeit a big one – was that we only have the right to own land *sufficient for our needs*; if we take more, said Locke, we are effectively robbing others, just as when we waste food, we offend 'against the common Law of Nature'.[42]

Locke Goes West

> . . . 'tis Labour indeed that puts the difference of value on everything.
>
> John Locke[43]

It's easy see why Locke went down so well across the pond. His vision of a peaceful agrarian society made up of industrious God-fearing folk reads like an eyewitness account of Protestant settler communities. Locke's belief in the duty to cultivate both chimed with settlers' values and helped to bolster their ambivalence towards Native Americans, who seemed to feel no such moral obligation. Even the mild-mannered Locke could barely conceal his contempt for the latter's seeming lack of initiative: 'I ask whether in the wild woods and uncultivated waste of America left to Nature, without any improvement, tillage or husbandry, a thousand acres will yield the wretched inhabitants as many conveniences of life as ten acres of equally fertile land doe in Devonshire, where they are well cultivated?'[44]

Locke's views would come back to haunt those who adopted them, not least those for whom his ideas seemed to offer a new vision of political freedom: 'Men being, as has been said, by Nature, all free, equal and independent, no one can be put out of this Estate, and subjected to the Political Power of another, without his own *Consent*.'[45]

If the words sound familiar, it's because they inspired the future author of the American Declaration of Independence, Thomas Jefferson. Born and raised on a plantation at Monticello, Virginia, Jefferson had a lifelong passion for agronomy, conducting experiments in his own garden on soil conservation, crop rotation and plant breeding. 'No occupation is so delightful to me as the culture of the earth,' he wrote, 'and no culture comparable to that of the garden.'[46] In Locke, Jefferson found a philosopher ideally matched to his political vision. Here was a social blueprint based not on a contract between subject and monarch but between independent, agrarian equals. When Jefferson drafted the historic Declaration, Locke's spirit guided him: 'We hold these truths to be self-evident, that all men are created equal, that they are endowed by their Creator with certain unalienable Rights, that among these are Life, Liberty and the pursuit of Happiness.'[47]

The words still have the power to move us, even though the society that they spawned hasn't turned out quite as Jefferson must have hoped. Indeed, as the developed world's second most *un*equal nation after Singapore, the United States mostly serves to illustrate how wide the gap between rhetoric and reality can get. In the nation that embraced Locke's egalitarian principles more fully than any other, the top 0.1 per cent now hold as much wealth as the bottom 90 per cent put together, and some 47 million people live in poverty.[48] For those failing to flourish in the 'Land of the Free', the theoretical right to pursue happiness must have rather a hollow ring.

What went wrong? In truth, the project was flawed from the start. Even as they were demanding independence from Britain, the colonists were busy snatching land from the Native Americans, an act of theft that, although it concerned Jefferson, he couldn't help but see from a Lockean perspective. He was similarly blind to the other flagrant injustice towards 'others', slavery. A slave owner himself, he did move to end the international slave trade in 1808, yet failed to ban the practice outright, and even brought his favourite domestic slaves from Monticello to the White House. All men are born equal, it seems, but some are born more equal than others.

There was another flaw in the system that Locke himself realised could prove fatal. As Aristotle had warned, human greed could be the greatest threat to democracy:

> Every Man should have as much as he could make use of . . . without straitning any body, since there is land enough in the world to suffice double the inhabitants had not the *Invention of Money*, and the tacit agreement of men to put a value on it, introduced (by consent) larger possessions, and a right to them.[49]

Long before the consumer age, both Aristotle and Locke worried that the pursuit of individual wealth could spell society's ruin. What neither could foresee, however, was the way in which money would come not merely to shape society, but to embody our very idea of the good.

The Gift

> To trade, the first condition was to be able to lay aside the spear.
>
> Marcel Mauss[50]

Money is so fundamental to our lives today that it can be hard to imagine a world without it. Everyday acts would take on a surreal quality (I'll give you a thousand chickens for that Honda), yet the quicksilver stuff that makes our world go round is far less reliable – and ancient – than we might think.

For most of human history, money didn't exist. Instead, as Marcel Mauss described in his 1950 book *The Gift*, societies were based on gift exchange. The Trobriand Islanders of Papua New Guinea, for example, sailed hundreds of miles in order to exchange shell bracelets and necklaces in a ritual known as the *Kula*.[51] Although ordinary exchanges of food and tools might take place alongside it, the *Kula* was not primarily a trading network as such, but rather an exchange of objects of great prestige that conferred honour and status on both giver and receiver.

The enormous investment of time and effort involved in the *Kula* can seem weird to modern sensibilities. Why sail hundreds of miles and risk drowning in order to exchange objects of no practical value that one doesn't even get to keep? The answer, Mauss argued, was that the gift exchange expressed the moral, spiritual and economic glue that bound societies together: 'Souls are mixed with things; things with souls. Lives are mingled together, and this is how, among persons and things so intermingled, each emerges from their own sphere and mixes together. This is precisely what contract and exchange are.'[52]

Despite the absence of cash, such gift economies depended, like ours, on high levels of trust. The exchange of gifts in such cultures was a grave and solemn act, with failure to reciprocate possibly leading to death or even war. In our digital age, it can be hard to imagine the burden of such obligations, although we occasionally feel an echo of them, such as when, for example, we're invited to a wedding and have to buy a gift for the happy couple. The angst that can attend our search for a suitable salad bowl or vase stems from the fact that, even in our material age, gifts possess something akin to a soul. Wedding lists may spare us the effort and potential embarrassment of giving, yet they also rob us of much of the pleasure. The recent trend for newlyweds to request gifts in cash takes the process to its logical, soulless conclusion.

Such complexities suggest one reason why money was so vital to the development of modernity. Abstract and impersonal, it took the agony out of exchange, relieving us of the rituals and obligations that once bound people together. Social bonds, although essential to our well-being, are antithetical to economic progress, getting in the way of its core goal of efficiency. Once people discovered what money could do for them, there was no looking back.

Paying in Clay

It should come as no surprise to learn that the origins of money lie in food. The fact that grain could be grown in surplus and easily stored and transported made it an ideal substance for early cities to trade. Almost as soon as they had worked out how to feed themselves, the Sumerian city-states of Uruk, Ur and Eridu were growing food for profit. By the third millennium BCE a trading network stretched all the way from their location in southern Mesopotamia to Syria and Anatolia, east towards Iran and south to the Persian Gulf and so on to India.[53] In exchange for wheat, the Sumerians imported copper, gemstones, lapis lazuli and alabaster with which to decorate their temples, homes and bodies. In the world's first cities, grain *was* wealth, and temple granaries were the reserve banks of their day.

The Sumerians didn't only grow grain, however; they also produced onions, garlic, peas, beans, lentils, cucumbers, lettuce, figs, dates, olives, grapes and pomegranates, as well as beef, mutton and pork and over fifty types of fish.[54] To trade such bounty, they needed a means of exchange more flexible than barter: they needed markets, and to make them function, they needed money.

The earliest forms of money were clay tablets mostly used to record agricultural deals such as the exchange of a cow for some barley. Since the deal could only be honoured once the harvest was in, the element of time entered the exchange, transforming what would once have been direct barter into an early form of forward contract.[55] Delaying the completion of a deal by the issuing of a credit note or IOU is essentially what money does, turning two people into a creditor and a debtor. Since the credit note is worthless unless the debtor honours it, the bond between lender and borrower is thus based on trust, hence the origin of the word credit (from the Latin *credo* – I believe).

A clay tablet representing a cow's worth of barley is not of course money as we know it today. The next stage in its evolution came when such tablets could be exchanged not just for barley but for anything equivalent in value to a cow – by anyone, any time, anywhere. Money, that is to say, morphed into an abstract token of value, whose worth depended more than ever on trust, which is why early coins tended to be made of precious metals, and why Julius Caesar had them stamped with an image of his head – so that citizens at the furthest reaches of the empire would still honour them.

Coins helped to establish trust and trade across vast distances, yet the fact that they were made of precious metals also caused confusion, since people began to associate the coins themselves with wealth, rather than with what they represented.[56] This error goes to the heart of the distinction made by Adam Smith between 'value in use' and 'value in exchange'. If you're dying of thirst, said Smith, a glass of water is more useful to you than a bag of diamonds: water, that is to say, has intrinsic worth – value *in use* – whereas diamonds have only value *in exchange*. Most of the time the relationship remains obscure – on an average day, for example, most of us would probably choose diamonds over water. It is only *in extremis* that the difference becomes clear. When the German Weimar Republic spiralled into debt after the First World War, for example, wheelbarrows full of banknotes were needed to buy a loaf of bread. While hyperinflation rendered money virtually worthless, the value of bread remained constant. Since food was what everyone needed, people reverted to the only form of exchange that works under any circumstances, barter.

Animal Spirits

> No one ever yet possessed so much silver as to want no more.
>
> Xenophon[57]

Money's capacity to spiral into oblivion has its origins in the great era of sea trade and exploration during the fifteenth century. In order to finance their voyages, the merchants of Venice needed to borrow large sums of money that they could only hope to pay back once their silks and spices had, with any luck, been safely landed and sold many months later. Due to the hazardous nature of seafaring, merchants paid lenders interest on their loans to compensate them for the risk they were taking. This created a further problem, however, since lending money at interest (usury) was banned by the Church. It therefore fell to the Jewish members of the community to act as moneylenders, a role that made them enviably wealthy. A new breed of financial service was born that took its name from the wooden benches – *banci* – from which the lenders conducted their business: banking.[58]

In the seventeenth century the Dutch became masters of the seas, largely through the creation in 1602 of the Vereenigde Oost-Indische Compagnie (United East India Company), the world's first public corporation. Funded partly by government and partly through the issue of public shares, the VOC behaved like a nation-state in its own right, negotiating treaties, establishing colonies and even waging wars in the name of trade. By mid-century, it had a virtual monopoly of trade in spices such as nutmeg, cloves and mace, with a lucrative sideline in Chinese gold and silk. Trade in shares was already so brisk by 1608 that a special building was constructed to house it, the Amsterdam Beurs, the world's first stock exchange. The Wisselbank or Amsterdam Exchange Bank opened the following year: the world's first central bank, it allowed merchants to trade directly from their accounts, either by direct transfer or cheque, much as we do today.

Although the buccaneering Dutch pioneered most of the instruments of capitalism, it fell to the Swedes to finish the job. Unlike the Wisselbank, which simply allowed merchants to shift deposits from one account to another, the Stockholms Banco began lending them out at interest. This practice, known as fractional reserve banking, is the cornerstone of modern finance.[59] As the name suggests, its principle is to allow banks to lend at interest, providing they keep enough in reserve (perhaps 10 per cent) to pay customers who might want to withdraw their cash. The beauty – and danger – of the system is that it vastly increases liquidity, which is to say that it puts money in circulation that has no basis in reality. With a bank operating a reserve of 10 per cent, a deposit of £100, for example, can generate a secondary loan of £90, which when deposited enables a further loan of £81, and so on. Within the space of just four transactions, the amount of cash swilling about has thus swollen from £100 to £271 (100+90+81), nearly three times the original amount.[60]

If you think this all sounds like one vast Ponzi scheme, you wouldn't be far wrong. Indeed, the system is designed to create a perpetual cycle of boom and bust in which investment in a confident, 'bullish' market sends share prices rising until they bear little relation to the underlying asset value, at which point people start to sell. If market euphoria has pushed prices too high, panic selling can set in, leading to the dreaded 'run on the banks' during which (unless they are deemed 'too big to fail') the banks can go bust. Markets, in short, respond to what John Maynard Keynes called animal spirits: mood swings which, when amplified by a flock mentality, can build into speculative bubbles that – as bubbles are wont to do – grow bigger until they eventually burst.

The Invisible Hand

> The great commerce of every civilized society, is that carried on
> between the inhabitants of the town and those of the country.
> Adam Smith[61]

Banking was always risky, yet by enabling trade in sugar, spices and slaves, it made seafaring nations like Britain and Holland unimaginably rich. By 1700, the wealth flooding into London and Amsterdam was not merely transforming those cities, but also their rural hinterlands, as farmers scrambled to meet the soaring demand for food. A raft of new agricultural practices pioneered mostly by the land-starved Dutch including crop rotation, cattle breeding and the growing of fodder crops came to Britain.[62] Marshes were drained and land enclosed and the countryside brimmed with entrepreneurial activity.[63] Poulterers and warreners took out loans to expand their businesses; fruiterers planted orchards and leased them out, while butchers became graziers and stockbreeders.[64] When Daniel Defoe toured the kingdom in 1720, he was in no doubt as to what was creating all the buzz:

> The country send up their corn, their malt, their cattle, their fowls, their coals, their fish, all to London, and London sends back spice, sugar, wine, drugs, cotton, linen, tobacco, and all foreign necessaries to the country ... London consumes all, circulates all, exports all, and at last pays for all; and this is trade.[65]

To Adam Smith, the contrast between all this entrepreneurial zeal and the relative torpor of rural France was puzzling. While staying in Paris in 1766, Smith met the leading economists François Quesnay and Jacques Turgot, whose group of 'physiocrats' (from the Greek *phusis*, nature + *kratia*, rule) was then grappling with the vexed question of how to feed Paris, which they hoped to do by modernising the medieval taxes and property laws that mired the rural economy.[66] All three economists agreed that the ultimate source of a nation's wealth was its land: Quesnay, indeed, went as far as to say that farmers were the only productive members of society, since landowners merely distributed the wealth they created, while merchants and artisans produced nothing at all.

While disagreeing on the latter point, Smith was impressed by Quesnay's proposal to remove all internal barriers to trade, a principle that Turgot tried to put into practice while serving as France's controller-general of finances in 1774. History might have turned out very differently had Turgot's tenure not happened to coincide with a series of disastrous harvests and consequent food shortages that were blamed, not on nature's intransigence, but on the physiocrats' meddling. Turgot was relieved of his duties, thus scuppering any chance of reforms that might have averted the 1780s food crisis that sparked the French Revolution.[67]

Meanwhile, the very different revolution unfolding in Britain led Smith down another path. While the physiocrats dreamed of turning France into a protectionist agrarian society, Smith imagined his homeland becoming something far grander: an industrial trading nation whose wealth would grow exponentially. The key, as he explained in a chapter of his 1776 *The Wealth of Nations* called 'On the Natural Progress of Opulence', was the trade that naturally took place between city and country:

> The town, in which there neither is nor can be any reproduction of substances, may very properly be said to gain its whole wealth and subsistence from the country. We must not, however upon this account, imagine that the gain of the town is the loss of the country. The gains of both are mutual and reciprocal, and the division of labour is in this, as in all other cases, advantageous.[68]

The division of labour was critical to the creation of wealth, said Smith, since it vastly increased efficiency. Using the famous example of a pin factory, he argued that specialisation (in which workers performed just one task) was the key to productivity.[69] But if everyone produced just one thing, how could they gain the necessities of life? The answer, said Smith, was by exchanging the product of their labour with that of others, via the market. Echoing Locke and prefiguring Marx, Smith argued that it was *labour* that should determine the exchange value of commodities; land, the ultimate source of wealth, came for free. The market would supply people's needs as if by an 'invisible hand', since natural self-interest would lead everyone to find their own niche within it: 'It is not from the benevolence of the butcher, the brewer or the baker that we expect our dinner, but from their regard to their own self-interest. We address ourselves, not to their humanity but to their self-love.'[70]

This last idea has proved controversial, since it seems to suggest that greed is good. Although Smith never put it that way, he did think that successful factory owners were good for society, since they would plough their profits back into their businesses, thus providing more work and so generating more wealth. This 'trickle-down' theory – the idea that all wealth is good, since it will find its way into the parts of society that other economies can't reach – is a central tenet of capitalism. Its flip side is the need for consumerism, since factory owners can't expand unless people buy more of their stuff. Fortunately, Smith observed, our appetite for the non-essentials of life was insatiable: 'The desire of food is limited in every man by the narrow capacity of the human stomach; but the desire of the conveniences and ornaments of building, dress, equipage, and household furniture, seems to have no limit or certain boundary.'[71]

In contrast to Aristotle and Locke, Smith saw boundless desire as the key to a good life, since it drove economic growth, the engine of universal prosperity. Far from being unproductive, therefore, merchants and artisans were the key to wealth generation, since they oiled the wheels of commerce upon which it was based. Instead of creating unhappiness, the 'natural progress of opulence' would create a world in which 'every man . . . becomes in some measure a merchant, and the society itself grows to be what is properly called a commercial society'.[72]

The 'Adam Smith Problem'

You don't need to have read *The Wealth of Nations* to be familiar with Smith's ideas, since they have become firmly embedded in the very way we think. Free-market capitalism enabled the growth of liberal democracy, so Locke and Smith can be considered the co-founders of Western modernity. Liberalism and capitalism are so fused in our minds today as to be virtually indistinguishable. Their principles are fundamental to our ideas of prosperity and freedom, so much so that we have barely noticed as *Homo sapiens* has morphed into *Homo economicus* and economic growth has become synonymous with our idea of the good.

Liberal democratic societies are among the happiest and freest in history, yet their success can't hide the fact that the 'natural progress of opulence' as promised by Smith has failed to materialise. While a minority have become ever richer, the majority stagnate in relative poverty. So what has gone wrong? Among those queueing up to answer that question, the general consensus seems to be that *Homo economicus* isn't human.[73] Ironically, this would not have been news to Smith. Indeed, had *The Wealth of Nations* not been such an immediate hit, Smith's fame would rest on his other great work, the 1759 *Theory of Moral Sentiments*, in which he noted the importance to human happiness of friendship and empathy:

How selfish soever man may be supposed, there are evidently some principles in his nature, which interest him in the fortune of others, and render their happiness necessary to him, though he derives nothing from it except the pleasure of seeing it.[74]

Contrary to what one might expect from the father of capitalism, Smith attributed much human misery to our misplaced 'disposition to admire, and almost to worship, the rich and the powerful, and to despise, or, at least, neglect, persons of poor and mean condition'.[75] Hardly the words of a man dedicated to the unbridled pursuit of wealth. Indeed, the vision of a good life in *Moral Sentiments* is positively Aristotelian, depending, not on the 'trinkets of frivolous utility' but rather on the capacity to reason, empathise and appreciate beauty.[76] Far from being based on greed, for Smith a good society depended on 'humanity, justice, generosity and public spirit'.[77] For him, in the end, it was all about love. 'Sympathy,' he wrote, '. . . enlivens joy and alleviates grief.'[78]

What, you may wonder, turned this cuddly Smith of 1759 into the 'greed is good' monster of 1776? The answer is: absolutely nothing. They were one and the same; indeed, Smith reworked *Moral Sentiments* until the end of his life, so the final version came out after *The Wealth of Nations*. The clue to the so-called Adam Smith problem lies in the fact that Smith's works describe two complementary halves of the world he knew. Writing at the dawn of the Industrial Revolution, he saw the benefits of mechanisation – his pin factory was based on a real example – yet never witnessed the devastation it would soon cause. He didn't live to see how the transformations necessary for industrialisation would destroy the very society that, for him, it existed in order to serve.

The Great Transformation

Social upheaval was essential for industrialisation because, as Karl Polanyi pointed out in *The Great Transformation*, the latter required the creation of a radically new social milieu, the 'market society'. Although markets had existed for centuries before the industrial age, the exchange of goods had been a predominantly social rather than an economic transaction in which trust, generosity and honour were paramount. The primary motive for such transactions, said Polanyi, was not profit but status. Smith's assumption that people would naturally take to life in a market economy was therefore false:

> No less a thinker than Adam Smith suggested that the division of labour in society was dependent on the existence of markets, or, as he put it, upon man's propensity to barter, truck and exchange one thing for another. This phrase was later to yield the concept of Economic Man. In retrospect it can be said that no misreading of the past ever proved more prophetic of the future.[79]

In reality, said Polanyi, life in a market society – in which the economy operated independently of social ties – was anything but natural. Yet such a market was necessary to industry, since the huge investment it required, such as for plant and machinery, would be too risky unless investors were guaranteed steady output and sales. Reliable supplies of raw materials and labour were therefore needed, and these could only be obtained through the commodification of the two key sources of wealth: nature and man:

> Machine production in a commercial society involves, in effect, no less a transformation than that of the natural and human substance of society into commodities. The conclusion, though weird, is inevitable; nothing less will serve the purpose: obviously, the dislocation caused by such devices must disjoint man's relationships and threaten his natural habitat with annihilation.[80]

The market economy required society to be destroyed, since its overriding need was for the market to be freed to operate without social constraints:

> To separate labour from other activities of life and to subject it to the laws of the market was to annihilate all organic forms of existence and to replace them by a different type of organisation, an atomistic and individualistic one ... In practice this meant that the non-contractual

organisations of kinship, neighbourhood, profession and creed were to be liquidated since they claimed the allegiance of the individual and thus restrained his freedom.[81]

Freedom in the new order meant the ability to engage in the market, an activity that would come to define Economic Man. As well as destroying landscapes and communities, therefore, the Great Transformation altered the values and meanings that had once underpinned people's lives. No society prior to that of Georgian England had conceived of such a thing as economic growth, let alone associated it with a good life. For most people, earning a decent living had been enough to be content; from now on, however, mere subsistence would not suffice. In the market society, one's primary goal in life had to be getting rich.

Early mill owners knew better than most how unnatural this new condition was. For those factory workers brought up on farms, the relentless monotony of their new work was hard to bear; most saw no point in enduring any more of it than necessary. When they had earned enough to live on for the week, therefore, most simply downed tools and went home. When the owners raised wages to encourage people to work longer hours, it had the opposite effect: the workers simply went home even earlier. Mill owners thus took the only alternative open to them, slashing wages to the point where workers could only survive by working all the hours available. A principle was established that remains central to capitalism: when starvation is the alternative, people will work for almost nothing.

Zero Hours

> All that is solid melts into air.
> Karl Marx and Friedrich Engels[82]

Those working today in Amazon fulfilment centres on zero hours contracts don't need to be reminded of where such logic has led. In a market economy, the ideal cost of labour is indeed zero. For Karl Marx, the problem lay in the fact that the 'means of production' (factories and machines) were exclusively owned by the bourgeoisie, who used their power to exploit the proletariat, forcing them to work for slave wages. In 1848 he and Friedrich Engels articulated this theory in their revolutionary anti-capitalist credo, the *Communist Manifesto*:

> The bourgeoisie, wherever it has got the upper hand, has put an end to all feudal, patriarchal, idyllic relations. It has pitilessly torn asunder the motley feudal ties that bound man to his 'natural superiors', and has left no other nexus between man and man than naked self-interest, than callous cash payment . . . It has resolved personal worth into exchange value, and in place of the numberless indefeasible chartered freedoms, has set up that single, unconscionable freedom – Free Trade.[83]

Marx and Engels recognised, as Smith had done, that capitalism depended on constant growth. Yet, far from representing the 'natural progress of opulence' to the communists, this simply meant 'everlasting uncertainty and agitation'.[84] Furthermore, the streamlining efficiencies necessary to create such growth robbed workers of all dignity and satisfaction: 'Owing to the extensive use of machinery and to division of labour, the work of the proletarians has lost all individual character, and, consequently, all charm for the workman.'[85]

It can be tempting, after reading the *Communist Manifesto*, to rush straight out and join the Party. Marx and Engels are persuasive critics, debunking capitalism with passion and precision. Yet their solution – that the proletariat should rise up and seize the means of production – has hardly been a storming success. On the contrary, their proposal that the state should control everything from commerce and communication to transport and agriculture has led to far worse suffering and oppression than anything liberalism has produced. Indeed both Russia and China have demonstrated that capitalism and communism can easily co-exist, segueing seamlessly into the former without relinquishing one iota of the state's grip on power. Even without the benefit of hindsight, the *Communist Manifesto* has a strikingly Hobbesian air: the communist state is, in effect, Leviathan by another name, and has proved more terrible than anything Hobbes imagined.

In most societies, the reality is that wealth and power tend to trickle up, not down. Whatever the underlying social vision, the gap between theory and reality is often yawningly wide. In despotic regimes this is something that the masses must either suck up or attempt to overthrow through revolution. But how can we explain the fact that such a divide also exists in the West?

Making Money

> Time is Money.
> Benjamin Franklin[86]

Two millennia before Gordon Gekko growled 'Lunch is for wimps', Aristotle warned of the dangers of the pursuit of wealth for its own sake. He would, one imagines, have found it rather ironic that our word economy is based on *oikonomia*, rather than on *chrematistike*, although the latter is admittedly rather hard to pronounce. Yet this conceptual slippage helps to explain how liberalism and capitalism merged to form the Leviathan that increasingly rules the West: neoliberal capitalism, or neoliberalism for short.

In his 1905 essay 'The Protestant Ethic and the "Spirit" of Capitalism', Max Weber traced the roots of this unholy alliance back to the Dutch Republic, in which Calvinists struggled to reconcile their belief in hard work and frugality with the cascade of goodies pouring in courtesy of the VOC.[87] Since Protestants believed that God put men on earth to be productive (John Locke, it should be noted, came from Calvinist stock), they concluded that, although the pursuit of wealth for its own sake was a sin, the accumulation of riches – providing one didn't *enjoy* them too much – was not wrong in itself. Indeed, the Protestant belief in predestination made many think that good fortune was merely a sign that God had chosen one for salvation.[88] The Dutch therefore redoubled their efforts to live pious, industrious lives in the midst of untold wealth, spending their cash not on luxurious trinkets but on a series of spectacularly funded good works.

For Weber, this capacity to remain sober in the face of phenomenal wealth was the essence of the capitalist spirit.[89] It was a talent that waves of Protestant migrants took with them as they sailed for the New World, where it proved the magic ingredient in that capitalist crucible. The spirit was captured by Benjamin Franklin in his 1748 essay 'Advice to a Young Tradesman, Written by an Old One', in which he spelled out the new entrepreneurial order:

> Remember that time is money. He that can earn ten shillings a day by his labor, and goes abroad, or sits idle, one half of that day, though he spends but sixpence during his diversion or idleness, ought not to reckon that the only expense, he has really spent, or rather thrown away, five shillings besides.[90]

Franklin, who made a fortune from his printing business and rose to become a signatory of the American Constitution, saw money chiefly as a liberator. He especially loved money's capacity to accumulate seemingly of its own accord: 'the more there is of it, the more it produces every turning, so that the profits rise quicker and quicker'.[91] Franklin remained true to his Puritan roots, however, advising his 'young tradesman' to remain modest and diligent, however much money he made:

> . . . the 'summum bonum' of this 'ethic' is the *making of money* and yet more money, coupled with a strict avoidance of all uninhibited enjoyment. Indeed, it is so completely devoid of all eudaemonistic, let alone hedonist, motives, so much purely thought of as an end *in itself*, that it appears as something wholly transcendent and irrational, beyond the 'happiness' or the benefit of the *individual*.[92]

While the absence of hedonism is hardly notable among today's super-rich, the compulsion to make 'money and yet more money' certainly is. Getting rich is now the dream of the masses, embodying the freedom to buy as many cars, yachts, houses and handbags as their celebrity idols. Capitalism may have turned its Puritan origins on their head, yet its central tenet that money is good in itself remains firmly entrenched.

Road to Serfdom

Socialism means slavery.
Friedrich Hayek[93]

The coming of the railways effectively completed the capitalist toolkit, unleashing the wave of corporate deregulation necessary to fund their construction. The UK Joint-Stock Companies Act of 1844 was a game-changing piece of legislation designed to ease the flow of capital in every way possible. Radically simplifying the process of incorporation, it allowed anyone to set up a company for a mere £10 stake and made corporations legal 'persons', due the same protections as private individuals. Limited liability was added in 1856, restricting investors' exposure to company debt to the value of their original stake.

The US soon followed the British lead, as various states competed for corporate investment by stripping away protections such as curbs on mergers and acquisitions, with the result that the number of US corporations plummeted from 1800 in 1898 to just 157 in 1904.[94] As railroad barons and industrialists made and spent vast fortunes and newly streamlined factories such as Ford's churned out cheap consumables (any colour you like as long as it's black), a period of economic euphoria – the Roaring Twenties – ensued. The American Dream seemed unstoppable until, on 29 October 1929, the unthinkable happened: the market crashed.

The immediate cause of the crash wasn't hard to spot: a classic speculative bubble fuelled by feverish market optimism. The more pressing question was how to solve the Great Depression that followed. For John Maynard Keynes, the answer was obvious. The economy had stalled, he said, because people had lost confidence in the market. Their 'animal spirits' were low and their instinct was therefore to hunker down and hoard money rather than spend it. The government therefore needed to restore confidence by injecting cash into the system and lower interest rates to boost the economy. Inspired by Keynes, Franklin D. Roosevelt's New Deal was a sweeping programme of banking reforms, benefits and infrastructure projects designed to put people back to work and create the closest thing that Americans had ever experienced to a welfare state.

Generally heralded as a great success, the New Deal didn't please everyone. To the Austrian-born economist Friedrich Hayek, it was a disaster. Hayek's personal experience of the rise of fascism during the 1920s and 30s (a consequence of the post-war financial meltdown) had given him a horror of state intervention in economic affairs. In his 1944 work *The Road to Serfdom*, he argued that such meddling could only lead towards state oppression and eventually totalitarianism. Socialism was akin to slavery, said Hayek, since it obliged everyone to fall in line and work towards the same goals. It was also practically inoperable, since it required the state to decide everything from the value of profits and wages to how wealth should be distributed. Such decisions, said Hayek, could never be made fairly, since society was far too complex for any individual to understand, let alone value. Only a free market – abstract, impartial and similarly complex – could be trusted to decide such matters. A free market, he argued, was the only guarantee of liberty, since it was the only mechanism that discovered true values and allowed people to decide how to act. 'We have,' he wrote, 'progressively abandoned the freedom in economic affairs without which personal and political freedom has never existed.'[95]

Thanks partly to its somewhat unlikely appearance in the April 1945 *Reader's Digest*, *The Road to Serfdom* found an immediate and enthusiastic audience in the US. Just as Locke had inspired a nation for whom liberty mattered above all else, Hayek appealed to those for whom equal opportunity meant the chance to get rich quick. All that was needed for his vision to become fully embedded in the American Dream – for liberalism to morph into neoliberalism – was for the US to transform itself, as it was already doing, from a nation of Lockean farmsteaders into an urban society.

Rise of the Plutocrats

> Wealth is the relentless enemy of understanding.
> J. K. Galbraith[96]

By the 1970s, the New Deal was all but played out. A series of world events including the Vietnam War, the 1973 oil crisis and the end of the

Bretton Woods system of monetary regulation was causing recession and social unrest on both sides of the Atlantic.[97] When Margaret Thatcher and Ronald Reagan came to power in 1979 and 1981 respectively, both saw socialism as the problem and Hayek as the solution. While Thatcher insisted there was 'no such thing as society' and Reagan joked that the scariest nine words in the English language were 'I'm from the government and I'm here to help,' both leaders set about deregulating banking, slashing tax, privatising public services and smashing trade unions. The neoliberal experiment had begun.

Four decades later, we know where it has led. Despite the 2008 crash – its most notable outcome to date – corporate power has continued to soar, creating levels of inequality not seen since the Victorian era. In 2018, America's top 350 CEOs earned 312 times more than their average employees' wage.[98] In both the US and UK, low and middle incomes have stagnated, while austerity measures have cut welfare to the bone. In the UK, 1.6 million food parcels were handed out in the year to April 2019.[99] Global trade, meanwhile, continues to serve corporate needs, with the World Bank, International Monetary Fund (IMF) and General Agreement on Trade and Tariffs (GATT) all forcing developing nations to remove 'barriers to business' such as environmental protections and workers' rights.[100]

It is some measure of neoliberalism's fatal allure that, in contrast to the reaction after the 1929 crash, the response to that of 2008 has been mostly to try to get back to business as usual. Despite being partly culpable for the calamity, British banks were deemed 'too big to fail' and bailed out to the tune of a cool £1.162 trillion.[101] And despite an ongoing procession of scandals including insider trading, offshore tax evasion and the illegal fixing of Libor (the inter-bank lending rate), financial regulation remains sketchy at best.

Only the maddest of Tea Party hatters could believe that neoliberalism delivers a good society, so why do we persist with it? The problem, as Joseph Stiglitz pointed out in *The Price of Inequality*, is that it works brilliantly for those at the very top. Contrary to Smith's labour-based model of value, most of the wealth in our economy is unearned. Rather than make pins for a living, the rich merely engage in various forms of 'rent-seeking': managing assets, earning interest and renting property. As

Stiglitz put it, 'Those at the top have learned how to suck out money from the rest in ways that the rest are hardly aware of.'[102] For the modern plutocracy, self-interest is precisely that and nothing more.

Since freedom in a market society depends on one's ability to pay, it doesn't take an economic genius to see that Hayek's theory doesn't stack up. On the contrary, as Michael Sandel noted in *What Money Can't Buy*, a free market guarantees that *only* those at the top enjoy real liberty.[103] Since the rich can afford to pay in order to see a doctor out of hours, jump a theatre queue or eat in a Michelin-starred restaurant, only they enjoy all that society has to offer. Furthermore, said Sandel, when one can pay to litter the countryside or to drive solo in a bus lane, money becomes a substitute for morality. In a market society, the richer you are, the less moral you need to be, an ironic outcome, given that the ideology insists that the rich must be hard-working and talented and the poor feckless or thick.

Nowhere are such beliefs more fervently held than in the US, where, despite all the evidence, wealth is still assumed to have been earned. Only in America could a presidential candidate get away with admitting, as Donald Trump did in 2016, to paying almost no tax on the basis that he was 'smart'. As the US economist J. K. Galbraith noted in his 1958 book *The Affluent Society*, Americans have a unique acceptance of others' wealth and a deep antipathy towards socialism, since they cherish the dream of one day becoming rich themselves, at which point they intend to cling to every hard-earned cent:

> The poor have generally been in favour of greater equality. In the United States this support has been tempered by the tendency of some of the poor to react sympathetically to the cries of pain of the rich over their taxes and of others to the hope that one day soon they might be rich themselves.[104]

Back in the *Happy Days* of the 1950s, most hard-working Americans *did* have a reasonable chance of making it big. Jobs were plentiful, the middle classes were in the ascendant and the American lifestyle was the envy of the world. Today in the US and UK, only the rich have much hope of getting richer. Perhaps the most remarkable aspect of the

American Dream is that, despite being decimated by globalisation, automation and corporate greed, it still endures.

Bring on the Robots

In 2016, British and American voters were given the chance to say what they thought of the neoliberal experiment so far. The results – in the shock form of the Brexit referendum outcome and the election of Donald Trump – sent a resounding raspberry to the liberal elites who had assumed that the votes were in the bag. Both results revealed deep anger among those whom Theresa May called the 'left behind': workers whose wages barely covered their living costs, who blamed their woes on globalisation, immigration and, above all, a corrupt political class. Both votes revealed deep social divisions that the neoliberal narrative had tried to paper over, prompting a furious debate between a hardening right and left – both claimed by newly emergent populist parties – about whether neoliberalism has finally run its course.

The rise of populism, which is also strong in nations such as Germany, France and Italy, raises questions that the traditional right and left aren't in the best place to solve. Industrial capitalism, as envisaged by Adam Smith, was an economic system based on the idea that making stuff generates wealth. Trade was beneficial to it, since it opened up new markets for producers and gave customers more choice. Since it was assumed that producers would plough their profits back into their businesses, more trade would naturally lead to more growth.

For a while the system worked fine – the new middle classes were evidence that the trickle-down principle actually worked. But with the financialisation of markets something strange began to happen: transactions took on a life of their own, seemingly disconnected from anything real. In 1890, the English economist Alfred Marshall responded by suggesting that, instead of being based on labour, the value of goods really depended on how much people were prepared to pay for them.[105] The wealth of nations rested, said Marshall, not on what people produced, but on what they *consumed*. This so-called neoclassical approach to economics offered a new way of measuring wealth, simply by counting what people spent and expressing the total as gross domestic product (GDP). Never mind that GDP ignored the real cost of manufacture and made no distinction between, say, money spent building bridges or the cost of clearing up after a crime; from now on, *any* economic activity would count as good.

Marshall's sleight of hand explains why nations like the UK and US remain near the top of the economic pecking order despite having slipped well down the manufacturing one. Our prosperity depends, not on producing food or widgets, but on financing deals based on goods made elsewhere. As is the case in most post-industrial societies, some 80 per cent of our economies are based on services (banking, transport, health, education and so on), meaning that a high proportion of our necessities are imported from abroad.[106] Although there is nothing wrong with this in principle, providing that the stuff we buy does no harm where it is made, the problem is that most of the blue-collar jobs that once underpinned our communities no longer exist, while those that do are threatened by cheap imports.

The dilemma that we face – between deregulation and protectionism – is essentially the same one which Adam Smith and the physiocrats grappled with over two centuries ago. The difference is that the social checks that provided a natural restraint on markets in the eighteenth century have since been destroyed by those very markets. The reason why Donald Trump's protectionist policies could never 'Make America Great Again' is that wealth these days stays where it is generated – in the financial system. The parallel tragedy of Brexit is that many of those who voted for it – people living in deprived ex-industrial regions – are precisely those whose livelihoods were destroyed by the free-market ideology of their Brexiteer champions.

In many ways, the crisis in the post-industrial world boils down to the difference between food and money. When food was the foundation of wealth, opportunities to get rich were limited, yet life was nevertheless grounded in material reality. However, our 5,000-year embrace of money has mirrored both the story of civilisation and the emergence of a kind of wealth that is profoundly disconnected from land and labour. The question that few are yet asking is what will happen when the industrial cycle goes full circle and those who currently produce our trainers and tiger prawns decide that they'd like to stop making stuff too, as is already starting to happen in China. Who is going to make our takeaway food and throwaway clothes then? Robots? And, if so, what are the ten billion or so people projected to be living on earth by 2050 going to do all day? Sext one another and play computer games?

Such questions reveal the real issue behind the populist movement: the problem of how to create a good post-industrial society. We might think that we already live in such a society, yet our consumerist life-styles are heavily subsidised by millions of unseen workers who suffer the same slave-like conditions that once prevailed in our own dark satanic mills. A 2013 report by the workers' rights group China Labor Watch found underage and pregnant workers in Apple's second-largest Chinese factory working sixty-six hours a week for less than half the living wage.[107] The following year, a *Guardian* report exposed human trafficking, forced labour and murder in the Thai prawn fishing industry – a situation that persists.[108] Such abuses aren't all far from home, either: a 2016 British parliamentary select committee reported that the clothing firm Sports Direct was run 'like a Victorian Workhouse', penalising workers for spending too long in the toilet or falling preg-nant and treating workers, in the report's words, 'as commodities rather than as human beings'.[109]

Treating workers like commodities is of course what capitalism has always done. Indeed, of all the things that a truly free market systemat-ically crushes, human dignity comes top. Today, it's not just weavers or farmers at risk either: recent developments in artificial intelligence mean that robots are already replacing doctors, lawyers and journalists, using pattern recognition to diagnose patients, parse legal documents and write copy. Leading researchers Carl Fray and Michael Osborne of the Oxford Martin School estimated in 2013 that 47 per cent of all American jobs were vulnerable to automation and that one third of white-collar jobs could be gone within a couple of decades.[110] Similarly, a 2017 study by the UK Office of National Statistics found that of 20 million jobs analysed, 1.5 million (7.5 per cent) were at high risk of automation, with a further 13 million (65 per cent) at medium risk.[111] Just as cotton mills once displaced farmers and craftsmen, the Second

Machine Age is poised to destroy the middle classes whose creation was arguably the first one's greatest – and some might say only – benefit.

Buddhist Economics

> One of the most fateful errors of our age is the belief that the problem of production has been solved.
>
> E. F. Schumacher[112]

Although automation will create new jobs as well as destroying old ones, there are unlikely to be enough to go round. So should we simply accept the inevitable and start building more bingo halls and golf courses? For some, the answer is an emphatic yes. In his 2017 book *Utopia for Realists*, the Dutch economist Rutger Bregman argues that we should embrace a new era of leisure in which people will work a maximum fifteen-hour week and live on the shared profits from the labour of mechanised slaves – robots – in the form of a universal basic income (UBI).[113] Such an outcome, Bregman argues, is 'what capitalism ought to have been striving for all along'.[114] For the first time in history, he says, we are rich enough to eradicate poverty by sharing the wealth that is, after all, merely the result of the 'blood, sweat and tears of past generations'.[115]

Could such a vision work? As a way of addressing capitalism's inequalities while removing the stigma of social welfare, UBI certainly has its attractions. Those in favour of it argue that its too-good-to-be-true simplicity would make it affordable, since it would do away with the Kafkaesque complexity of modern welfare systems. But what of the human side? Would we be able to cope with a life of boundless leisure, or would we – as some retirees already do – simply lapse into purposeless stupor? The question is far from new, indeed it was posed as long ago as 1930 by none other than John Maynard Keynes. Believing that progress would lead inevitably to a shorter working week, Keynes worried that people would struggle to cope with lives of endless leisure, a problem he saw as 'the greatest challenge for future generations'.[116] Since Keynes considered meaningful work essential to a good life, he argued that the transition to a leisured society would need to be gradual, as people invented new ways of occupying themselves: 'For many ages to come the old Adam will be so strong in us that everybody will need to do some work if he is to be contented. We shall do more things for ourselves than is usual with the rich to-day, only too glad to have small duties and tasks and routines.'[117]

It might not be time to take up crocheting quite yet, however. Contrary to Keynes's predictions, we are actually working more, not less. So what is going on? We already know the answer for gig workers: like their Victorian factory forebears, they are forced to work long hours just to make a living. But what about the managerial class? One 2013 study by the Harvard Business School found that managers and professionals across Europe, Asia and the US were spending between eighty and ninety hours per week either working or monitoring work on their phones.[118] The short answer is that, in a competitive world, doing crazy hours at work is one way of staying ahead, creating a workaholic culture in which doing overtime is tacitly expected.

For the Anglo-German economist E. F. Schumacher, the problem was not so much how long people worked, but the nature of the work itself. In his 1973 book *Small Is Beautiful*, he argued that industrial workplaces were robbing people of their humanity, leading to rising depression, drug use and crime. 'Is it not evident,' he wrote, 'that our current methods of production are already eating into the very substance of industrial man?'[119] The problem, said Schumacher, was the way in which capitalism split human activity into production and consumption. While all pleasures were excluded from work in the name of efficiency, workers were expected to compensate by spending their wages on pleasure in their leisure hours: 'If man-as-producer travels first-class or uses a luxurious car, this is called a waste of money; but if the same man in his other incarnation of man-as-consumer does the same, this is called a sign of a high standard of life.'[120]

The absurdity of this, said Schumacher, is that 'man-the-producer' and 'man-the-consumer' are *one and the same person*, whose happiness could be just as easily secured at work as at home. He cited an industrial farmer who admitted that he wouldn't dream of eating his own food and felt fortunate to be able to buy organic produce grown 'without poisons' instead. When asked why he didn't simply grow organic food himself, the farmer replied that he 'couldn't afford it'.[121]

Equally damagingly, said Schumacher, our economic confusion means that we fail to value our greatest source of wealth, nature. Because we didn't make it, we treat it as though it were free, yet the fact that we didn't produce it ought to make us value it more, not less. In truth, we should treat nature as beyond price – 'meta-economic' – as though it were sacred.

What, Schumacher wondered, might an economy that valued land and labour look like? An economy, for example, based on Buddhist beliefs? Since Buddhists revere nature, such an economy would aim to minimise consumption, helping to preserve the natural world. Because Buddhism also recognises the importance of good work in nourishing the human spirit, such an economy would also seek to make work more fulfilling, not to dehumanise it. 'It is clear, therefore, that Buddhist economics must be very different from the economics of modern materialism, since the Buddhist sees the essence of civilisation not in a multiplication of wants but in the proliferation of human character. Character, at the same time, is formed principally by a man's work.'[122]

A Buddhist economy would thus lead to a life lived more in harmony with nature, in which craftsmanship and care played a far greater role. Rather than leaving the work to robots, it would promote mindfulness and productivity as well as recognise our need to live within ecological bounds.

Sitopian Economics

> Only a crisis – actual or perceived – produces real change.
> Milton Friedman[123]

Half a century after Schumacher's economic thought experiment, its relevance has only grown. Our need to exist within planetary boundaries is clearer than ever, raising multiple questions about how we might live good lives in the future. As a natural counterbalance to neoliberalism, Buddhism is a useful place to start; but could we learn to live according to its principles? Could we give up our consumerist, individualistic lives for more mindful, collaborative ones?

The answer partly depends on your view of humanity: whether, like Aristotle and Locke, you think we're capable of leading balanced, mutually cooperative lives, or whether, like Hobbes and Hayek, you think we're not. In truth, the reality probably lies somewhere in between. What we know, however, is that we're far better at collaborating in times of crisis. It's no accident that the most visionary social programmes to come out of the US and UK in the past century – the New Deal and the welfare state – came in the wake of the Wall Street Crash and Second World War respectively. When we experience common hardship, we naturally pull together, becoming more empathetic, altruistic and visionary. Crises make us realise how precious our everyday lives are to us, so for a time at least, we appreciate what we already have. Crises, in short, give us the chance to readjust our values, which is why our failure to change course after the 2008 crash may turn out to be the greatest missed opportunity of the century.

Our failure to act was partly down to the fact that we had no obvious Plan B ready to swing into action. This makes it doubly urgent that we come up with one now. We know that we're in a mess because we've forgotten the true value of things: we've been living, as Schumacher noted, on natural capital. It's clear that we need a new economy geared towards helping us flourish within our ecological means, one that we might therefore usefully base on food. As the one substance that we must all consume every day, food occupies a unique place in our world. Consisting of living things from nature that we consume in order to stay alive, it has an intrinsic sacred value of the sort Schumacher described. Food, quite literally, *is* life. If we don't value it, we deserve to be doomed.

You might think that we already value food – an organic chicken, after all, costs at least double the price of its industrial cousin. Yet the price disparity reflects far more than just the different ways in which the birds were raised. When we buy hand-reared organic produce, it seems expensive because it reflects the true cost of producing food that is good in every sense: nutritious and tasty as well as ethically and eco-logically produced. The trouble is that this is the only sort of food that reflects its true cost. The other sort – the industrial food that supplies more than 95 per cent of our diet – is artificially cheap due to the sys-tematic externalisation (often through government subsidy) of the true cost of producing it.[124]

Many of the costs of industrial food – deforestation, soil erosion, water depletion, exhausted fish stocks, pollution, biodiversity loss, rural depopulation, unemployment, obesity, chronic disease, climate change and mass extinction – aren't counted in the price we pay in the shops. Although it is hard to put a price on such externalities, they are far from trivial. A 2017 report by the Sustainable Food Trust, founded by British organic farmer Patrick Holden, found that we pay twice for our food in the UK: roughly £120 billion in the shops, and the same again in hidden external costs.[125] If we did true cost accounting for food, Holden argues, we would not only damage ourselves and the planet far less, but would end up paying less to sustain ourselves overall.

In reality, 'cheap food' is an oxymoron – an illusion created by indus-trial producers and governments keen to disguise the true cost of living. While externalities such as deforestation, pollution and climate change are accounted for elsewhere, industrial farmers who would otherwise struggle to make a living from the low prices we pay for food are sub-sidised by the state. So what might the world look like if, instead, we were to internalise the true cost of our food? The answer is that indus-trial farming would rapidly become unaffordable, while ecologically produced organic food would emerge as the bargain it has always been. Buying food would become a virtuous circle, in which the market would favour foods that nurtured nature, animals and people. Ethical, ecological producers would sell more produce, so gaining some of the economies of scale now reserved for industrial agri-food giants. Schumacher's farmer could afford to grow organic food as well as eat it.

Such an approach is what I call sitopian economics, a value system based on the intrinsic worth of food. Since food affects virtually every aspect of our lives, adopting such an economy would have an immediate, even revolutionary effect. By changing the ways in which we produce, transport, trade, cook, share and value food, we could transform our landscapes, cities, homes, workplaces, social lives and ecological footprint. Valuing food is about far more than enjoying an organic carrot or nice piece of cheese; it's our only path to survival. The good news is that by valuing food we also get to eat healthily and deliciously, what you might call a win–win scenario. As well as being something we must consume every day, food is our most reliable source of pleasure. By bringing economy and pleasure together, we can create a gastro-economic philosopher's stone.

Sitopianism is based on the simple premise that by valuing food we value life. Indeed, Buddhism and sitopianism share many common values: respect for nature, social justice, good work, mindfulness and the avoidance of needless violence. By letting food be our guide, we can create a practical, adaptable, universal Plan B. Good food, as Epicurus noted, is the fulcrum of a meaningful, happy life. If we all eat well in the future, the odds are that we'll live well too.

Root and Branch

> In cases of need, all things are common property.
> Thomas Aquinas[126]

One stumbling block to creating a sitopian economy is the fact that we've got used to spending so little on food. In 1950, the British spent between 30 and 50 per cent of their income on it; today we spend just 9 per cent (the least in Europe), compared to 13 per cent in France and 25 per cent in India. Americans, meanwhile, spend just 6.4 per cent, the lowest figure anywhere in the world. This is not to say that American food is cheap to produce, however; on the contrary, at an average annual cost of $2,273 per head, it is the world's most expensive, costing more than ten times what it costs to feed the average Indian.[127]

Affordability is a key reason why food is an area where politicians fear to tread. We can't raise prices, the argument goes, since it would push millions into hunger. Yet surely the problem is the other way around? Shouldn't we be asking why, in some of the richest nations on earth, millions are living below the breadline? The real taboo here isn't food, but our refusal to face the dark heart of neoliberalism. The irony is that the solution lies in the problem: by valuing food, we could create many good jobs that could lift people out of poverty. Food and farming remain by far the largest employers on earth, providing work which, when valued, can be hugely rewarding. For numerous reasons – stewardship of the land, maintaining biodiversity, boosting rural communities and simply eating better – we need *more* people working in food, not fewer. If we were to value food again, everyone – even the fattest of cats – would benefit.

Another common objection to paying for high-quality organic food is that the small-scale farms from which it tends to come mean going back to the past. Yet nobody said we have to do all this without the aid of technology. On the contrary, new technologies are already helping us work with nature, rather than against it. Computational agriculture, for example, is an emergent field that uses robots not to replace humans but to help them farm more naturally. Computerised sensors combined with precision drone mapping, for example, can help farmers to monitor the moisture and mineral content of the soil, telling them exactly which fields and plants need their attention. The idea that humans and technology are mutually exclusive is a false polarity created by capitalist logic. What we need is new forms of partnership between them. Indeed, when land is scarce and humans abundant, it would be crazy not to employ as many people in food as we possibly can.

Fortunately, we don't have to wait for governments to act in order to effect positive change through food. People all over the world are already doing it for a range of reasons: to protect local resources, maintain ancient traditions, fight globalisation, or simply through the desire to eat better and connect more with nature. In the West, such initiatives can take the form of organic box schemes, community kitchens and gardens, microbreweries and bakeries, food co-ops and community-supported agriculture (CSA) projects, in which city-dwellers pay farmers in advance to grow their food for them and even come to help out on the farm. In developing nations, such groups include peasant and small farmers' co-operatives, community cooking and growing groups, small producers and indigenous people fighting to keep their land and water rights.

Under the umbrella of international movements such as Slow Food and Via Campesina (the international peasant and small farmers' network), these groups represent the global food movement, the scale and scope of which becomes clear when thousands gather every two years in Turin for the Terra Madre, the Slow Food festival held in the vast, aircraft-hangar-sized old Fiat factory. Farmers and producers from Egyptian bakers and Ethiopian honey producers to Mexican chilli-pepper growers and Russian dairy farmers come together with cooks, writers and activists from all over the world to swap food, knowledge, projects and politics. All have stories to tell about fighting the forces of globalisation and trying to preserve local traditions that may be hundreds if not thousands of years old.

The food movement, one soon discovers, is about much more than food. It is democracy in action. No matter whether they are peasants fighting for their land rights or wealthy urbanites fed up with eating junk food, those who perceive food's true value share a common belief: that the right to eat as one chooses is fundamental to freedom.

Slow Money

> Food should be good, clean and fair.
> Carlo Petrini[128]

The connection between food and freedom highlights the greatest barrier to creating a better sitopia: power. If one were to draw the industrial food system as a diagram, it would resemble a tree, with millions of producer roots and consumer branches joined by a single agri-food trunk. Such is the stranglehold of corporate Big Ag on the global food system that just four companies – ADM, Bunge, Cargill and Dreyfus – control more than 75 per cent of the worldwide trade in grain.[129] These agri-food giants use their power to push commercial crops like corn and soy onto local farmers at the expense of native produce. Despite widespread concern about monopolies, mergers continue apace. In 2018, the

German seed company Bayer won approval to buy rival Monsanto for a cool $62.5 billion, creating a company (nicknamed Baysanto) with more than a quarter share of the global seed and pesticides market.

Once one has grasped that food systems and societies reflect one another directly, it becomes obvious that a democratic food system can't look like a tree. As our ancestors knew very well, control of food is power, a basic truth that we seem to have forgotten. If we want a free society – a democratic global village – it follows that we need a different food system: one characterised not by monopoly but connectivity. This idea goes back to the roots of liberalism and Locke's agrarian vision, which was based on the principle of food sovereignty. If we want to live in a democracy, we need to take back control of our food.

Slow Food founder Carlo Petrini believes we can do this by becoming 'co-producers'. Eating well, says Petrini, is about more than just pleasure; it is a social responsibility akin to that of being Rousseau's good citizen. Since we must all eat, we have a moral duty not to be mere consumers, expecting our food to arrive as if on a magic carpet. Instead, we should get actively engaged in feeding ourselves, at the very least by appreciating what it takes. This needn't involve digging potatoes or milking cows at four in the morning; it could simply mean supporting a CSA or local farmers' market or joining a box scheme – directly connecting the producer roots and the consumer branches. In essence, it means channelling our attention, appreciation and hard-earned cash towards those who make good food with love.

One group already doing this is the US-based non-profit organisation Slow Money. Founded in 2010 by the venture capitalist and Slow Food enthusiast Woody Tasch, the group supports small-scale quality producers by investing, as Tasch puts it, 'as if food, farms, and fertility mattered'.[130] By 2019, Slow Money had invested over $66 million in 697 organic farms and food enterprises in the US, Canada and France, a drop in the ocean compared to the billions invested in industrial food, yet a significant counter-current nonetheless.

For Tasch, Slow Money is not just about food; it is also a response to the fast money of the digital age. During the course of 1960, he points out, three million trades were made on the New York Stock Exchange; today the annual figure is closer to five billion.[131] 'If money is speeding around so fast,' he asks, 'should we be surprised that we don't know what the hell it's doing?' The digital era, Tasch believes, demands new social and fiscal policies to 'bring money down to earth'. We must, he says, find ways to invest 'for the benefit of future generations'.[132]

Like its parent Slow Food, Slow Money envisions a future in which good food will go mainstream. Is such a vision possible or merely elitist? Slow Food has, after all, been accused in the past of being all about gourmandism and fine dining. Yet, as Carlo Petrini is at pains to point out, the reality is quite the opposite: good food is not about vast expense, he insists, but about the care and skill with which it is made. Far from being elitist, the rural traditions from which such food comes are bastions of resistance against the real elitism in our world, that of free-market capitalism. Indeed, before globalisation started wiping them out, such rural traditions were as common as the proverbial muck in which the food was grown.

Community Chest

Slow Money offers a glimpse of how sitopian economics might work. To create a truly sitopian society, however, would require major shifts in government policy, beyond food and farming to urban and regional planning, trade, taxation, transport, health, education and energy. By harnessing food's potential to transform each of these sectors – for example by prioritising food and farming in regional planning decisions, or by procuring local organic food for hospitals, prisons and schools – governments could generate more employment, boost health and wellbeing, build stronger communities, create a more beautiful environment, reduce carbon emissions, protect biodiversity and improve food security.

One example of how this might work is the Preston Model. In 2011, the depressed Lancashire city was dealt a severe blow when proposals to revitalise its historic centre with a £700 million mall development anchored by John Lewis and Marks & Spencer (following the developer-led, retail-based pattern adopted by numerous British cities since the 1990s) collapsed.[133] Bereft of inward investment, councillor Matthew Brown realised, the city would need to find ways of building its wealth from within. Inspired by examples of new economic development including the Mondragón cooperatives of northern Spain, Brown approached local public institutions such as colleges, museums, schools and housing associations, asking them to redirect as much of their expenditure as possible towards local suppliers.[134] In 2013, such bodies were spending hundreds of millions on contracts for construction, maintenance and meal provision, yet just one pound for each twenty spent stayed in Preston. It was this economic leakage that Brown sought to reverse.

Working with the Manchester-based Centre for Local Economic Strategies (CLES), Brown went out of his way to involve public service providers, for example breaking down the 2015 Lancashire County Council tender for school meals into nine different lots to supply yoghurt, eggs, cheese, milk and so on, so that the contracts were small enough for local businesses to handle. The approach worked: local suppliers using Lancashire farmers won every contract, boosting the local economy by an estimated £2 million.[135] Results elsewhere were similarly spectacular. In 2013, six of the institutions approached by Brown had been spending £38 million in Preston and £292 million in Lancashire as a whole; by 2017 these figures had leaped to £111 million and £486 million respectively. As the *Guardian* writer Aditya Chakrabortty has noted, the awarding of local contracts through such 'guerrilla localism' has a multiplier effect, as cash recirculates through the local economy, creating jobs that in turn lead to more spending on goods and services, generating ever-greater employment.

From its 2011 nadir, by 2018 Preston had risen, phoenix-like, to be voted the UK's most improved city in the Good Growth for Cities index.[136] After declaring itself the first living-wage employer in northern England in 2012, it set about systematically reversing its economic downward spiral, with measures including the backing of a local credit union to take on the loan sharks. In 2019, Preston launched a programme to incubate ten new worker-owned cooperative businesses, providing premises rent-free for a starter period to get the firms on their feet. Now council leader, Brown also plans to create a new cooperative bank for north-west England, focused on lending to small businesses in the region.

As such initiatives suggest, the Preston Model is about much more than just boosting the local economy; it is also about restoring the sense of ownership, togetherness and pride that come with belonging to a thriving community. Perhaps the proudest sign of Preston's resurgence is opposite the town hall in the city's historic marketplace, where instead of a faceless, off-the-peg retail monolith funnelling cash out of the city, there stands the award-winning old covered market, lovingly restored by local firms in a £4 million contract and once again selling regional produce to local residents. One could hardly wish for a better symbol of the engagement, participation and *oikonomia* that are the true basis of community wealth.

Preston has shown how communities can take back control of their economies in order to rebuild local democracy and resilience. The potential pay-offs of scaling up such initiatives are vast, yet for such an approach to go mainstream would require political vision and international cooperation. We need these anyway, of course – no community or nation in the modern world can act in isolation. Meaningful change will necessarily demand new structures of global finance and governance in order to shift power away from bodies such as the IMF and World Bank and their corporate clients.

Food can play a key role in the formation of such new political alliances. Those calling for urgent action on climate change, for example, such as Greta Thunberg and the school climate strikers, Extinction Rebellion and campaigners for the US Green New Deal, are creating a platform for a new, transnational, ecologically based politics. In 2019, the former Greek finance minister Yanis Varoufakis launched the Democracy in Europe Party, with the aim of using a Green New Deal to unite progressives across Europe, just as populist movements have united around immigration and nationalism. Financed by green investment bonds issued by the European Investment Bank (EIB), the party's policy would focus on the transformation of food and agriculture, energy, housing and manufacture to create sustainable, secure jobs for the future.

Letting Go

> The market and private property should be the slaves of democracy.
> Thomas Piketty[137]

As the 2015 Paris Agreement on climate change demonstrated, shared concern for our planet can unite us. The success of initiatives such as Fair Trade have shown that rich and poor nations can trade with one another without the latter being bankrupted. However, as Thomas Piketty argued in *Capital in the Twenty-First Century*, before we can truly act together, we need to implement a vast international exercise, forgiving debt and redistributing wealth.[138] Such an exercise, for Piketty, is essential not just for social justice but for future global stability. If left unchecked, he believes, the current economic system will lead to 'terrifying consequences', creating a dysfunctional, unstable world. We need to stop thinking of socialism and capitalism as irreconcilable opposites, he argues, and make them work together for the common good.

Such ideas are far from new. The ancient Hebrews, for example, were a wandering people for whom social cohesion was far more important than riches. Rather than base their economy on accumulation, therefore, they used it for redistribution. Fundamental to this was the tradition of the Sabbath year, during which the land was to be rested:

> For six years sow your fields, and for six years prune your vineyards and gather their crops. But in the seventh year the land is to have a Sabbath of rest, a Sabbath to the Lord. Do not sow your fields or prune your vineyards. Do not reap what grows of itself or harvest the grapes of your untended vines. The land is to have a year of rest.[139]

Every forty-nine years (every seventh Sabbath year) a Jubilee year was held, in which all debts were forgiven and all rented land returned to its owners. Since forty-nine years was around the average lifespan of two generations, the Jubilee allowed the children of debtors to regain their land, while the land itself was given a chance to recover.[140]

The Jubilee is far from unique: among traditional peoples some kind of periodic reckoning, often aligned to earthly cycles, is ancient and universal. The potlatch, for example, was an annual winter festival during which, having spent all summer hunting, fishing and generally stockpiling food and other goods, Alaskan and British Columbian tribes gathered for a gift exchange much like the *Kula*. The added twist, as Marcel Mauss described, was that once exchanged, the gifts were extravagantly destroyed:

> In a certain number of cases, it is not even a question of giving and returning gifts, but of destroying, so as not to give the slightest hint of desiring your gift to be reciprocated. Whole boxes of candlefish oil or whale oil are burnt, as are houses and thousands of blankets. The most valuable copper objects are broken and thrown into the water.[141]

The potlatch, Marcel Mauss noted, was a kind of ritual cleansing, allowing tribes to enjoy the pleasures of material wealth without succumbing to the spell of possession. In following the rhythm of the seasons, the potlatch echoed the natural ebb and flow of earthly riches, attuning the human experience of growth and decay to planetary cycles. It was only with the onset of industrialisation that the concept of permanent growth would take hold, and the commandment to rest would become the one that humans would most consistently flout.

A Sitopian Contract

From ancient clay tablets to the digital flicker of bitcoin, money's evolution has mirrored our own. Like a fickle god, it has enabled us to trade, collaborate, invest and prosper, yet it has arguably destroyed as many dreams as it has created. Fused in our minds with freedom and happiness, it has imprisoned us in a system that is both divisive and destructive.

If we are to free ourselves from money's grasp, we need a way of thinking that transcends it; an economy based on values grounded in reality. Food can give us this. Since it embodies what we all share – the struggle to lead good lives – it can form the basis of a new social contract between us all. Imagine yourself sitting down to a wonderful dinner with friends and family, the people you love most in the world. It's a special occasion and everyone has helped to produce the meal, bringing dishes, cooking together, laying the table and so on. There's plenty to go round and everyone helps one another before helping themselves. Drinks are poured, conversation flows, and a warm sense of bonhomie descends. You feel happy and loved and at peace with the world.

Now imagine that you are sitting down to such a meal, but with millions of strangers. How would you feel? What would you serve and how would you make sure that everyone got enough to eat? The idea may sound absurd, but the truth is that we *do* eat together, every day of our lives, with every living creature on earth. We just don't do it around the same table. If we *did* share our meals directly, we'd probably pool our resources very differently, extending instinctive hospitality to include all our fellow human and non-human diners.

When it comes to food, we are natural sharers – and, furthermore, our instinct scales up: it is axiomatic among caterers, for example, that when cooking for 500 people, one allows less per head than when catering for fifty, or five. When we eat together, we have an inbuilt sense of fairness, taking less than we might otherwise to make sure there's enough to go round. Such instincts go back to the sociability and altruism of tribal life. Sharing food prompts our sense of oneness, while sharing money just makes us greedy and jealous. Unlike cash, food satisfies us because it unites us, not least because we intuitively know there's only so much we can store.

This was another founding principle of liberalism. John Locke assumed that if nobody took more land than they needed, there would be plenty to go round. He is yet to be proven wrong. In the end, the redistribution of wealth must come back to land. A sitopian contract would be based on the right to food sovereignty for all species, with everything that implies.[142] It would seek to end ecological destruction, monopoly and slavery and to establish the right of every human and non-human to eat well, now and in the future. Through food, it would seek to build resilient, collaborative networks, just as, millennia ago, human societies first evolved.

A New Kind of Growth

We must cultivate our garden.

Voltaire[143]

Would living in a good sitopia make us happier? Clearly not all of us are sufficiently obsessed with food to want to spend all day thinking about it, which is absolutely fine. But such people might still enjoy the flourishing countryside or vibrant town centres that a sitopian economy would foster. As the UK Countryside Agency pointed out with its 2002 'Eat the View' campaign, the landscapes we love are often the product of organic, mixed farming. Similarly, anyone who has strolled through a market square has experienced food's power to shape and animate space. Public life in our towns and cities revolved around food and could be revitalised by a resurgent food culture, as has already happened with recent trends towards quality street food in cities like London and Portland, Oregon. Even those with extreme gastro-ennui can't help but be affected by the way we eat.

Whether or not we're foodies, valuing food will be vital to all our futures, since it is key to living within our ecological bounds. As we move towards a zero-carbon economy, we're going to need a new concept of economic growth that doesn't imply endless expansion and destruction and instead mimics nature. We need to create what the American economist Herman Daly calls a steady-state economy: one that fits within planetary boundaries.[144] As Daly explains in his 1996 book *Beyond Growth*, the term sustainable growth is an oxymoron, since it implies that growth can be limitless. The idea is inherited from classical economists, who took raw materials for granted – Adam Smith, as we saw, assumed that wealth from land was effectively free. Since we now know that natural wealth is limited, Daly argues, we need a new sort of economy based, not on infinite growth, but on a balanced form of sustainable development that stays within natural limits. Instead of producing ever-greater quantities of things, such a steady-state economy would instead focus on increasing quality.

Daly's insight is critical to how we can lead good lives in the future, since the potential to increase quality is virtually limitless. As the experience of those living in low-carbon communities such as BedZed suggests, post-industrial life can offer opportunities for flourishing that three centuries of industrialisation have largely obscured. With more time, space and opportunity to develop our skills and imagination, we can redirect our attention away from consumerist distractions and back towards our own development. We can forge new social connections and friendships and evolve a closer relationship with nature that doesn't increase our footprint on the globe – all of which brings us back to food.

A good sitopia is a naturally zero-carbon society, since all food comes from nature, and exemplary farming practices nurture and mimic natural ecological cycles. Food-centred economies are thus the natural basis of resilient communities, since they put the *oikonomia* back into economics. Living in more locally productive, interactive societies would make us healthier and fitter. The creation of local farms and gardens would result in more beautiful, greener environments, which would be beneficial to our wellbeing.[145] We'd be less worried about climate change and would live in a fairer world, something also known to be central to happiness.[146] Last but not least, we would regain some of the agency over our lives that the digital era has progressively eroded.

Growing food is one of those things that you don't really get until you try it. As a flat-dwelling Londoner who had barely grown a thing in her life until a few years ago, my recent excursion into growing Danish pickling cucumbers, or *asier* (introduced to me by my good friend, chef Trine Hahnemann), has been little short of revelatory. Gardening is the opposite of consumerism: one must be active, engaged, patient, observant, empathetic and, above all, in synch with nature. No wonder the garden is a universal symbol for paradise.[147] Gardens make us naturally happy, yet they also stand for everything we've lost on our way to civilisation. We may no longer live in the primal landscapes of our ancestors, yet our tended plots, however small, remain the expression of everything upon which we still depend. If the shared meal is a metaphor for a good society, the garden stands for the economy upon which it rests.

The point, as Smith, Locke and others pointed out, is that our economy really *is* a garden. The natural world is the source of all our wealth, enriched by those who tend it, so if we want to thrive, we need to start thinking more like gardeners. If we were planning to cultivate a garden, what would we want it to look like and how might we plan to nurture it? We would clearly want it to be rich and diverse enough to nourish us and our descendants. We would need to tend it carefully, learning which plants thrive in certain places and which happily coexist, cutting back where necessary to encourage new growth. We'd make compost to enrich the soil and foster as much foliage as possible. Our garden would grow, not by spreading, but by gaining inner richness.

By valuing food, we can nurture, rather than destroy, our commonality. By feeding ourselves and others well, we can build on instincts buried deep in the social fibre of our being. Feeding people is one of life's greatest pleasures and privileges – a good society might be built around the motto 'Feed thy neighbour as thyself.' Food can guide us through uncertain times if we put it back where it belongs – at the heart of society, where, in reality, it has always been.

5
City and Country

Brooklyn Grange

I'm on a farm. It's not a big one – perhaps one and a half acres in total – yet it produces seventy-plus varieties of organic fruit and vegetables, plus eggs, honey and flowers. It's early October and production is in full swing. On sale at the stall are rocket and kale, purple and white radishes, aubergines, bushy-topped carrots and some lethal-looking red and yellow chillies, while stretching away in neatly raised beds are tomatoes, cabbage, lettuce, chard, cucumbers, beans, peas and herbs, all topped off by a row of tall golden sunflowers.

So far, so normal. What makes this farm different, however, is what lies beyond the sunflowers: the unmistakeable skyline of New York City. It's rather a good view, actually, since, unlike your typical farm, this one is twelve storeys up, on the roof of an old shipbuilding facility in the heart of Brooklyn Navy Yard. Welcome to Brooklyn Grange, the world's largest soil-based rooftop farm.

I'm chatting with Ben Flanner, the farm's co-founder, president and director of agriculture. Tall and wiry, with an affable grin that spreads easily under his floppy hat, Flanner has the air of one who has found his true vocation in life. Born and raised in Wisconsin, he spent most of his childhood outdoors, playing games and gardening with his mother. 'I did have a lot of fun always in the vegetable garden,' he recalls. 'I transplanted all these raspberries from my grandma's house and put them in our backyard, and when we moved I brought them with us – that was always kind of my project.'[1]

After gaining a degree in industrial engineering, Flanner moved to New York and began a promising career in management consultancy. After a year, he was posted to Australia to do a cost analysis on a winery, where he found himself increasingly envious of the workers. 'I guess that was the start of my curiosity,' he says, 'I just felt cooped up in this office, which was next to the nursery building. A lot of people were coming through all day long with dirt on their hands. They'd been out in the sun and they were thirsty for water and hungry for lunch, and they sure as heck looked a lot happier than I was.'[2]

When he returned to New York, Flanner began visiting farms whenever he could. He was attracted to farming, he realised, but he also loved the city, with all its energy, creativity and sociability. Was there some way, he wondered, that he could have both? There was no eureka moment, he says, just a gradual realisation that urban farming might be the answer. 'It was one of those few times in my entire life when I've not had much hesitation,' he recalls. 'I said I'm going to leave by next spring, and I'm going to go into agriculture.' The next problem that Flanner faced was finding somewhere to farm. One of the densest cities on earth, New York isn't exactly flush with opportunities for would-be agrarians. The breakthrough came when he was leafing through *New York* magazine and saw a beautiful wildflower meadow that, he was amazed to discover, was several storeys up. He realised he could farm on a roof. He got straight on the phone to Chris and Lisa Goode, whose meadow it was, who found him a 6,000-square-foot warehouse in Brooklyn owned by film producer Tony Argento, who could see the potential of a rooftop farm and so was willing to pay for the installation of a green roof membrane and tonnes of a special lightweight soil called Rooflite, while Flanner and co-farmer Annie Novak planned what to grow.

In the spring of 2009, Eagle Street Rooftop Farm – New York City's first such commercial farm – opened for business. It was an immediate hit. Locals loved having fresh fruit and veg on their doorstep, while the press went mad for two photogenic young farmers squatting among their cabbages in front of the Manhattan skyline. Flanner and Novak were soon selling all the produce they could grow, and Flanner realised that he needed to expand. After much searching, with two new partners on board, he opened Brooklyn Grange in Queen's in 2010, followed by Brooklyn Navy Yard in 2012.[3]

Today, Brooklyn Grange grows 50,000 pounds of organic produce per year, selling to local markets, shops and restaurants and directly to the public from its rooftop stalls. With twelve full-time staff and thirty seasonal workers, the farm has a community-supported agriculture (CSA) scheme as well as an educational programme which has so far welcomed 25,000 youngsters through its doors. Flanner, meanwhile, has become something of a celebrity, in high demand as a green roof and urban farming consultant. He remains passionate about food's power to build healthy communities, yet he is more down to earth (albeit from twelve storeys up) than he once was about rooftop farming's potential to feed cities.

Surveying the skyline, Flanner reflects: 'I would look across and say "Man, *every one* of these roofs should be covered with food!"' Today he knows better. While covering flat roofs with vegetation makes perfect sense in most cases, improving water retention, providing urban cooling and enhancing roof longevity, to farm on a roof, Flanner has found, is a very different proposition. 'You really do need those perfect roofs,' he says. 'We've learned a whole lot about how the whole system works, the price of food, the efficiency of it, the physical rigour of going up and down. I would never put a farm on another roof unless it had a freight elevator, and not that many buildings have a freight that comes to the roof.'

The Vertical Farm

Flanner thus puts the kibosh on the idea, cherished by many, of turning concrete jungles into productive paradises. While urban farms bring a refreshing slice of rurality into the city, they are never going to be able to feed cities in their entirety. This is due to what I call the urban paradox: the fact that cities, whose main function is sociability, have always relied on somewhere else – the countryside – to feed them. If cities were to include all the farmland that produces their food, which in London's case is an area roughly a hundred times bigger than the city itself, they would cease to be cities as we know them and would instead become some sort of urban–rural hybrid.[4] The paradox is that, although those of us who live in cities think of ourselves as urban, in reality we still dwell on the land.

For some, however, the dream of cities feeding themselves is far from dead. On the contrary, vertical farming – growing food indoors in cities – has a committed and expanding following. The idea was first conceived by US microbiologist Dickson Despommier as he drove past some empty offices on his way to work. We grow food in greenhouses, he thought, so why can't we stack greenhouses on top of each other in cities? Not only would doing so protect crops from weather and pests; it would release vast tracts of farmland back to grassland and forest to resume their natural 'ecosystem services and functions'.[5]

After an interview published in *New York* magazine in 2007, Despommier was bombarded with proposals from architects and designers keen to make his idea a reality.[6] Despommier would amuse himself by sitting at his desk, putting a tick or cross beside each increasingly weird and wacky design. No vertical farm yet existed, but the buzz was growing as technical advances in light-emitting diodes (necessary to replace sunlight) brought the idea ever closer. Dutch researchers found that plants grew happily under a low-energy spectrum of blue, red and far-red light – they can't absorb green, which is why they look verdant to us – and thrived on specially tailored nutrient recipes delivered either hydroponically (via water) or aeroponically (via mist).[7] Both systems were highly water-efficient: hydroponics were found to use 70 per cent less water than conventional horticulture, while aeroponics used 70 per cent less again.

Back in NYC, an agricultural professor by the name of Ed Harwood had been tinkering in his shed to produce the perfect aeroponics growing mat (made of recycled plastic) and the finest of spray nozzles. When Columbia Business School graduates David Rosenberg and Marc Oshima came across Harwood's work, they knew they had struck vertical farming gold. The three set up a company called Aerofarms and investment was soon flooding in. By 2016, backers including Prudential Financial and Goldman Sachs had ponied up $50 million between them: even at New York prices, that's an awful lot of salad leaves.

That same year, Aerofarms opened its flagship farm in a 70,000-square-foot disused steel mill in Newark, New Jersey. The largest indoor farm yet built, it is capable of producing two million pounds of microgreens (baby kale, watercress, pea shoots, etc.) per year, grown in tray-like beds stacked a metre apart and up to twelve layers high, like bunk beds on some giant troopship. The plants grow under controlled-spectrum

LED lighting, their roots dangling in nutrient-rich spray, and are closely monitored to determine their optimal growing algorithm, which, as the firm's website proclaims, allows the team to control their 'size, shape, texture, color, flavor and nutrition with razor-sharp precision'.[8]

With year-round production, multiple growing layers and up to 30 annual harvests, Aerofarm claims to have yields 130 times greater than those of a conventional farm.[9] All this without soil, sun, rain, pesticides, tractors or even a weather-beaten face in sight – all vertical farmers wear hairnets and lab coats and have the pasty look of those who rarely see sunlight. Although pipped to the post as the world's first commercial vertical farm, an honour claimed by Singapore's Sky Greens in 2012, Aerofarm's self-styled global headquarters certainly signals the scale of the firm's ambition. With plans to expand across the US and beyond, CEO David Rosenberg makes no secret of his mission to 'transform agriculture around the world'.[10]

Rocket Man

> The future is already here – it's just not evenly distributed.
>
> William Gibson[11]

Could vertical farms really feed cities in the future? Their advantages are certainly compelling: year-round food production, an end to weather-related crop failures, no pesticides or herbicides, reduced food miles, water conservation, minimal waste, urban employment and, last but not least, delicious baby leaves plucked just hours before, available in a shop or restaurant near you.[12] What's not to like?

The answer is absolutely nothing – as far as they go. As a way of growing salad leaves in or close to cities, vertical farms stack up, as those already thriving in cities including New York, Singapore and London prove. Yet, enthusiastic though they are, most vertical farmers are realistic about the limits of their enterprise. London's first vertical farm, Growing Underground, whose name stems from the fact that it occupies a subterranean network of air-raid shelters thirty metres beneath Clapham Common, was an overnight sensation when it opened in

2015, sparking huge media interest, striking deals with the likes of Waitrose and enticing Michelin-starred chef Michel Roux Junior to become a non-executive director.[13] Despite their success, however, co-founders Richard Ballard and Steven Dring remain level-headed about the scalability of their leafy empire. 'Vertical farms are part of the answer to feeding cities,' says Ballard, 'but they will never feed them in their entirety.'[14] There's the rub: man cannot live on rocket alone, leaving open the question of how the vast bulk of our food will be grown in future.

Despommier is typically upbeat on the prospects, arguing that cities could grow up to 80 per cent of their food in block-sized buildings on urban peripheries.[15] While acknowledging that cows and sheep probably don't belong on such farms, he sees no reason why everything else we eat – including pigs and chickens – couldn't. There is plenty of precedent for this: pigs and chickens have cohabited with humans for thousands of years, feeding on kitchen scraps and even – in the case of one nineteenth-century piggery in the London Borough of Kensington – sleeping under their beds.[16] The question is whether we think it's ethical to keep such naturally forest-dwelling animals indoors – assuming, that is, that we intend to carry on eating bacon.[17]

Pigs might fly in the vertical farming future, but what of the millions of tonnes of cereals and pulses needed to keep them and us fed? We currently cultivate 1.6 billion hectares of farmland, an area equivalent in size to South America, and keep an area roughly double that size under permanent pasture, which is a heck of a lot of farm to bring indoors.[18] Apart from the question of where such farms would be built, there is the matter of cost: Google, never one to give up easily on a potential money-spinner, abandoned its Alphabet X vertical grain-farming project in 2015, on the basis that it couldn't work out how to grow cereals effectively (i.e. profitably) indoors.[19] A glance at current prices suggests why: while microgreens retail at a healthy £3.70 per 100 grams, a tonne of wheat costs in the region of £150, making grain 247 times less valuable by weight, not quite such an attractive investment for the likes of Goldman Sachs.[20] Similarly, while local baby leaves have the cachet of freshness and deliciousness, there is no equivalent added value to grain grown on one's doorstep.

Next comes the question of ownership. As with all high-tech ventures, the steep costs of setting up vertical farms demand big returns for investors. However altruistic it may sound, vertical farming is still a business. Despommier, like the true utopian he is, has suggested that governments might fund the research necessary to get vertical farming truly off the ground. In reality, however, such research – and the patents to go with it – is already in private hands: as Aerofarms CEO David Rosenberg told the *New Yorker* in 2017, 'We are so far above everybody else in this technology, it will take years for the rest of the world to catch up.'[21]

The most compelling reason why vertical farming isn't the answer to feeding cities, however, is the matter of what the plants are grown in. Like most conventionally produced fruit and veg, they rely on a solution of chemicals (NPK and other minerals) that is effectively the plant equivalent of Soylent. Leaving aside the question of whether lettuces grown in such chemicals are as nutritious as their soil-based counterparts, the claim that they are produced in the city is therefore only partially true. Vertical farming doesn't solve the urban paradox, since it relies on fertilisers – the food that our food eats – imported from elsewhere.

Like all utopian ideas, vertical farming seems miraculous until you read the small print. While it will no doubt play a role in the way we feed our cities in the future – not least in oil-rich desert states or land-starved Singapore – it is unlikely to replace much of the 49 million square kilometres of land that we currently farm. Which is a shame, because the issues that it seeks to address are very real indeed.

Urban Paradox

Over the 5,500 years in which cities have existed, the urban paradox has dogged their citizens. The inconvenient truth that cities need countryside – and vice versa – has been evident right from the start: 'countryside' is in fact an urban construct, a space outside the city that exists at the latter's behest. Although our ancestors gardened for thousands of years before they started building cities, the irrigated fields, groves, gardens and orchards of ancient Sumer were of a different order. The first landscapes made explicitly to serve urban populations, they were in effect the world's first countryside. The question so beloved of archaeologists – whether cities or farming came first – doesn't really matter in the end: agriculture and urbanity co-evolved, and out of their union sprang urban civilisation.[22]

Living in a modern city, it can be easy to forget how powerfully geography shaped our past. Industrialisation rendered the bond between city and country all but invisible, yet for most of history it governed everyday life. Most cities were built on rivers, not just as a source of fresh water and fish and a handy way of disposing of waste, but also as a means of transport. Access to rivers and the sea was vital to pre-industrial cities, as we have seen, since it was far easier to carry heavy goods like grain by water than overland, limiting the size to which land-locked cities like Paris could grow.[23] Other foods, however, were produced closer to home: many households (not just in Kensington) kept pigs or chickens, and all cities were surrounded by market gardens, where fruit and vegetables could benefit from generous doses of night soil – animal and human waste – carefully gathered for use as manure.

Sumerian city-states pioneered the principles of self-sufficiency that the Greeks would later admire and that, for many, remain the ideal even today. Situated on the fertile floodplains of the Tigris and Euphrates, they built massive irrigation systems to tame and channel the flood waters and so render the landscape fertile. Covering a staggering 10,000 square miles by 1800 BCE, this artificial landscape was considered an extension of the civilised urban realm.[24] The world's first countryside thus emerged as a liminal territory somewhere between city and nature, tame and wild, civilised and uncivilised; a position it has occupied ever since. Yet the most significant result of this division of land into urban and rural was to create a duality at the heart of civilisation.

What did the world's first city-dwellers make of urban life? Did they think it was a good thing, or did they pine for a simpler existence closer to nature? Although we'll never know the answers to these questions, a remarkable text – the 4,000-year-old *Epic of Gilgamesh*, the oldest story in existence – at least gives us some hints. The story opens with the eponymous hero Gilgamesh, the ruthless young ruler of Uruk, pitilessly driving his people to build a wall around the city. Tired of being mistreated by their king, the people appeal for help to the gods, who decide to curb Gilgamesh's excesses by creating a wild rival for him, Enkidu, equal in stature and strength and thus able to quell the raging 'storm of his heart'.[25] A wild creature with matted hair who grazes on grass, Enkidu is inducted into the ways of the city by the harlot Shamhat, who after seducing him offers him bread and ale – the foods of the city – before shaving and anointing him with oil.[26]

Enkidu comes to the city to challenge Gilgamesh to a fight, which ends with them becoming sworn friends. All is well until Gilgamesh insists, against Enkidu's advice, that they set off to kill Humbaba, the guardian of the sacred cedar forest. The pair find Humbaba and, also against Enkidu's advice, Gilgamesh slays him and proceeds to fell the sacred trees. Enkidu gathers up some of the timber to build a new door for the city temple, but his attempt to appease the gods fails and he dies. Heartbroken, Gilgamesh travels to the end of the world – donning the skins of wild beasts and becoming semi-wild himself – to seek flood survivor Uta-napishti, who supposedly possesses the secret of eternal life. Uta-napishti, however, informs him that all men are mortal and that he must learn to accept the fact. Broken and dejected, Gilgamesh returns home, yet upon seeing his city again, he realises that its walls will outlast him and thus will be his monument.

Life and death, food and sex, love and rivalry, loss and redemption: the *Epic of Gilgamesh* has it all. At its root, however, is the recognition that living in cities comes at a cost. Like Enkidu, we have to *learn* to be civilised, yet in making the transition from nature to culture we risk losing our roots in the wild. Enkidu is Gilgamesh's alter ego: neither good nor bad, his wildness represents what Gilgamesh lacks – the natural counterpart to urban life necessary to achieve balance. Bonding with Enkidu is essential to Gilgamesh's path to maturity. This process not only requires Enkidu's taming but his sacrifice, which prompts Gilgamesh's own journey into wilderness, the practical and symbolic reverse of Enkidu's earlier one. Finally Gilgamesh is made to face his own mortality. Only then can he return home, self-aware and complete at last.

What is remarkable about the *Epic of Gilgamesh* is how powerfully it describes our dilemma as civilised beings. The archetypal hero's journey, it reveals how the urban paradox affects us, not just politically and economically, but as individuals. Finding a balance between city and country is vital, it suggests, not just so that civilisation can function, but so that we can find a balance between our wild and civilised sides. If we withdraw into cities entirely, it warns, we risk becoming sick. Out of kilter within ourselves, we lose respect for our fellow humans and for nature (as Gilgamesh does by enslaving his people and despoiling the sacred cedar grove), losing our sense of place in the world. To be truly civilised, it suggests, we must remain grounded in nature.

Utopia

As well as being a rattling good yarn, the *Epic of Gilgamesh* could also be seen as our oldest work of political philosophy. Scattered among its tales of gods, beasts, harlots and heroes is a powerful meditation on the nature of civilisation. The product of the world's first urban society, it raises questions that each subsequent one has been bound to ask. What is remarkable is how little, over the four millennia since the story was written, those questions have changed.

What does it mean to be civilised? What must we sacrifice in order to attain such a state? Given our dualistic natures, how should we live? Such questions are all part of the urban paradox as well as central to utopianism. Gilgamesh's two defining acts – building a wall and destroying the forest – express our deepest physical dilemmas. In order to create, we must destroy; our task, therefore, must be to learn to balance our needs with those of nature.

For Plato and Aristotle, the key was to keep cities small. Plato's ideal city (*polis*) described in his *Laws* contained 5,040 'hearth fires' – households led by male citizens – suggesting a community of 30,000–35,000, once women and slaves were included.[27] His city was surrounded by farmland, divided in such a way as to give each household two plots, one in the city and one in the countryside. Although Plato didn't mention *oikonomia* directly (probably because it was obvious), it is clearly the principle upon which his city is based. In order to keep numbers constant, he proposed shipping off surplus citizens to found new colonies elsewhere and sharing out 'spare' sons between families – the social implications of which he left conveniently vague. While acknowledging that self-sufficiency was desirable for the *polis*, Plato's main aim in keeping it small was to create a community in which everyone knew each other, for there could be 'no such boon for a society as this familiar knowledge of citizen by citizen'.[28]

For the more practical Aristotle, self-sufficiency was the overwhelming priority. Since the ultimate goal of the *polis* was independence, it could not afford to rely on others for food. *Oikonomia* was thus the key to political freedom, and the scale necessary to achieve self-sufficiency the determining factor in the *polis'* ideal size: 'a household is a more self-sufficient thing than the individual, the state than the household; and the moment the association comes to comprise enough people to be self-sufficient, effectively we have a state'.[29]

Since the *polis* would ideally supply all its own needs, the division of labour was also vital to its success, yet once it was big enough to do everything necessary, there was little to be gained from it growing any larger, since this would merely stretch its resources and make it harder to defend. Although democratic Athens had some 30,000 citizens and thus a population of some 250,000, most Greek cities were indeed much smaller: few had populations above 30,000 and many limited their growth by doing just as Plato advised, dispatching 'volunteers' to go off and found new colonies elsewhere.

When it came to building cities in the ancient world, small was generally beautiful: the city-state model was so successful that for the first three millennia it was the only show in town. Around 750 BCE, however, a city was founded that would buck the trend so spectacularly that it would turn the very idea of urbanity on its head. That city was ancient Rome.

Consumption City

Even today, the scale of ancient Rome can take your breath away. Many of the classical city's 2000-year-old ruins still stand, their vast tufa blocks doing sturdy service, whether in the rotund form of Hadrian's Pantheon or lining the walls of the Cloaca Maxima, the city's main sewer. With a million citizens at its height, Rome was arguably the most audacious city ever built. Matched in size before nineteenth-century London only by a handful of medieval Chinese capitals, it was the world's first true metropolis. So how on earth did it manage to feed itself?

The short answer is: with difficulty. As we've just seen, transporting food in the ancient world was at least as great a challenge as growing it: one only has to imagine how much grain would have to have been trundled into Rome on ox carts to realise that the city could never have fed itself from its local hinterland. The key to its success was that it had access to the sea. At its height, Rome was extracting everything from grain, oil, wine, ham, salt, honey and *liquamen* (fermented fish sauce) from all over the Mediterranean, Black Sea and North Atlantic.[30] Alexandrian grain ships were the supertankers of their day, a lifeline for Rome's hungry citizens, up to one third of whom relied on a free grain dole handed out by the state. If you thought food miles were a modern phenomenon, think again.[31]

Rome was the blueprint of a new kind of city based not on self-sufficiency or trade, but consumption.[32] Much like the UK today, its food strategy relied on imports, an approach that affected not just foreign suppliers but also local farmers. With the bulk of its food coming from abroad, local producers were free to concentrate on highly lucrative villa farming (*pastio villatica*), growing fruit and vegetables and producing other delicacies such as milk-fed snails and nut-stuffed dormice for the luxury market, specialising much as vertical farms and organic box schemes do today.[33] While local producers thrived, however, Rome's distant hinterlands suffered: by the third century AD, the soils of North Africa were exhausted, and observers wrote despairingly of the white, caked earth, a sure sign of fatal salinisation.[34]

By relentlessly extracting its nutrients from distant lands, Rome effectively ate itself to death, yet it was far from the first or last great civilisation to do so. Indeed, the pattern has been remarkably consistent. The Sumerians, whose genius for irrigation wasn't matched by an equal talent for drainage, met a similar fate.[35] The Greeks' obsession with self-sufficiency didn't help them much either: their clearance of forests during the fourth century BCE in order to grow more wheat on steep fragile slopes led to widespread soil erosion. Plato described the Attic hills as 'the skeleton of a body wasted by disease'.[36]

Feeding cities has never been easy. Yet, despite their agricultural mishaps, our urban ancestors held the land in deep respect. Cultivation and culture, as we saw earlier, were powerfully connected in the Roman mind. Like the Greeks and Sumerians before them, the Romans considered cultivated land (*ager*) a rural extension of *civitas*, the civilised realm of the city.[37] Wilderness, in contrast, was viewed with disdain bordering on dread, the home of unruly barbarians who were the very opposite of civilised life.[38]

It took a city the size of Rome to reveal the true extent of the urban paradox. At one extreme, we can live close to nature ('separated and scattered', as Tacitus said of the Germans), eschewing the benefits of civilisation. At the other, we can live in great metropolises, suffering the stresses of urban life in exchange for its opportunities.[39] Or we can adopt a third way somewhere in between, living in small towns or cities with close links to the countryside – or vice versa – which for most of history is what most people have done.

A unique representation of this third way is Ambrogio Lorenzetti's 1338 fresco, the *Allegory of the Effects of Good Government*, which occupies one wall of Siena's great council chamber, the Sala dei Nove.[40] A richly detailed evocation of daily life in the medieval city-state, it shows a bustling city and similarly industrious countryside (the fields and groves can still be glimpsed out of the chamber's window) whose mutual prosperity is based on cooperation and trade. The fresco illustrates just what it says on the label: how the 'effects of good government' are a perfect symbiosis between city and country. Had Siena's councillors glanced up at the image during one of their meetings, its message would have been clear: look after your countryside, and it will look after you.

After gazing at the Lorenzetti for a while, a nagging thought creeps in: Why, if its message is so clear, aren't there more paintings like it? Surely every town hall in the world should have something similar on its walls? The short answer is that the medieval Italian city-state, like the Greek *polis*, was a rare example of a society in which something like an urban– rural balance *was* reached.[41] Most cities in history have not worked in harmony with their hinterlands, but have instead chosen to exploit them. Although close ties between city and country were the norm in the pre-industrial world, power-sharing was not. As the *Allegory* makes clear, if you want to achieve harmony between city and country, you must make it the core of your politics.

Goodbye Geography

Few of us today can see the landscapes that feed us from our window. As we saw earlier, the arrival of the railways in 1825 transformed the way cities were fed, making it possible for the first time to build them more or less any size, shape or place.[42] The effects of this transformation were soon felt in Britain: in 1800, just 17 per cent of the population lived in cities – an unusually high number, due to the Industrial Revolution – yet by 1891 the figure had leaped to 54 per cent, making Britain the first urban-industrial nation on earth.[43]

As the metropolitan carpet began to spread in the UK, a matching agricultural one was rolled out in America's Great West, where swathes of previously inaccessible prairie (cohabited by Native Americans and millions of bison) were linked to the East Coast for the first time. The 1830 opening of the Baltimore and Ohio Railroad signalled an era of economic expansion and ecological destruction on a scale never before seen. First to go were the bison, slaughtered for their hides or just for sport from moving trains, a massacre so relentless that the southern herd was wiped out within just four years.[44] With the bison gone, their human companions soon followed, scattered or removed to reservations, leaving the plains and prairies open for conversion to grain production.

All railroads led to Chicago, whose strategic position on Lake Michigan close to the Mississippi watershed put it in pole position to profit from the new trade. As East Coast and European cities clamoured for food, Chicago was ideally placed to deliver. With mountains of grain pouring in, Chicagoans had to figure out what to do with all the surplus, and the idea they came up with was to feed it to cows. By 1870, Chicago's Union Stockyards formed a square-mile city within a city, employing

75,000 people and processing three million head of cattle a year. As William Cronon noted in *Nature's Metropolis*, such numbers presented the meatpackers with a problem: the issue was not so much how to kill the beasts (which was done with ruthless efficiency) but rather 'what to do with the animals once they were dead'.[45] Up until now, meatpacking (feeding animals on grain and then packing them in salted grain for export) had focused on hogs, and for good reason: sausages, bacon and ham were popular ways in which swine flesh could be preserved long after slaughter.[46] Beef was a different story, however, since most Americans liked to eat their steak fresh. Cattle generally arrived 'on the hoof' for this reason, before being slaughtered by local butchers. If the packers wanted a slice of the action, they had to get smart.

The man who cracked the problem was Gustavus F. Swift, one of Chicago's biggest packers, who had more incentive than most to get his dressed beef (butchered carcasses) to the East Coast in an edible state. Swift's idea was to mount blocks of salted lake ice on both ends of each railway truck, so that chilled air flowed over the beef and kept it fresh. With ice-houses spaced along his route, he was soon able to sell his beef in Boston, nearly 1,000 miles away. What Swift had invented was the chill-chain: refrigerated delivery routes that were the last piece of the food logistics puzzle. With aggressive marketing and cut-throat prices, Chicago meatpackers soon persuaded Bostonians and New Yorkers that factory beef slaughtered hundreds of miles away was better than fresh meat from their local butchers. Industrial food – and cheap meat – had arrived.

The Chicago meatpackers were the founding fathers of the modern food industry. With their efficiencies of scale, logistical mastery and ruthless business practices, they drew the parameters that still define industrial food. By controlling the entire supply chain (thus achieving the holy grail of so-called vertical integration) the packers gained unprecedented power, which they used relentlessly to undercut rivals and drive them out of business. By 1889, just four companies controlled 90 per cent of the Chicago beef trade and thus most of the US supply.[47] Today, a greater oligopoly including some of the original Big Four controls the global market: acquired in 2007 for $1.5 billion by the Brazilian giant JBS, Swift now forms part of JBS-Swift, the world's largest meat processor.[48]

As global demand for meat continues to soar, remote landscapes are being transformed to satisfy it. In 2018, 7,900 square kilometres of Brazilian forest – an area five times the size of London – were cleared to make way for cattle grazing and soy production, the latter mostly for animal feed.[49] Just as Americans once destroyed their natural hinterland in the name of profit, Brazilians are destroying theirs, and, thanks to the Big Ag cartel, Americans are still reaping the rewards. The widespread burning in 2017 of forest in the *cerrado* – an area of wooded savanna one fifth the size of Brazil – was found to be linked to soybean production for Cargill and Bunge, key suppliers to Burger King.[50] JBS-Swift, meanwhile, has been linked to the destruction of Serra Ricardo Franco State Park, a primordial Amazonian forest thought to have inspired Arthur Conan Doyle's *Lost World*, in order to graze 240,000 head of cattle.[51] In 2018, newly elected far-right President Jair Bolsonaro raised further fears when he announced his intention to merge Brazil's ministries of environment and agriculture, putting agribusiness in charge. By July 2019, concern that the president had given the green light to illegal logging was confirmed when the government's own monitors announced that rates of deforestation had increased by 88 per cent since the previous year and were now at a record high, with an area the size of Manhattan being cleared every day – figures that Bolsonaro denounced as 'lies'.[52]

No landscape today is sacred. Out of sight and mind, even the wildest terrains – including those deep under the sea – are being transformed by what William Cronon calls the 'geography of capital', an approach to nature in which only the profits to be made from it count.[53] The ancient dance of city and country has mutated into a final, fatal land-grab.

Changing Places

How we feed cities matters because the future looks set to be overwhelmingly urban. According to the UN, 54.5 per cent of us now live in cities, and the figure is expected to rise to 68 per cent by 2050.[54] One fifth of us (1.7 billion) live in cities with populations of over one million, of which thirty-one are classified as megacities, metropolises of ten million inhabitants or more. Of these, the five largest are Tokyo, Delhi,

Shanghai, Mumbai and São Paulo, with populations of 38, 26, 24, 21 and 21 million respectively; all except Tokyo are projected to grow by between 24 and 38 per cent by 2030.[55] Unsurprisingly, given the scale and speed of such growth, many cities are struggling to cope, so that an estimated one billion people worldwide live in informal shanty towns, or slums.

What is causing this headlong rush into cities? Is it a good thing? And if not, what if anything can we do about it? Whatever the causes, the fact is that, for most people today, urbanisation and progress are synonymous. Nowhere is this truer than in China, which for three decades has been pursuing the most radical national urbanisation programme ever seen. Over the past quarter-century, 500 million Chinese have moved into cities: think Industrial Revolution, but ten times quicker and one hundred times the scale.[56] The effects of this massive upheaval have been felt around the world. So what can they tell us about the relative benefits of urban and rural life?

The ancient promise of urbanity was to free humans – some, at least – from the shackles of subsistence to follow higher pursuits. The problem was that, from the start, such benefits weren't equally shared. The scribes and poets of Sumer who wrote the *Epic of Gilgamesh* were not those engaged with an ox and plough.[57] Farming enabled civilisation, yet it also divided society into two mutually dependent communities whose outlooks were worlds apart. The partnership between city and country was always to some extent antagonistic.

We still live with the effects of that schism – even after five millennia, we can't see across the divide. City-dwellers have long thought themselves superior to their country bumpkin cousins; even Friedrich Engels thought that industrialisation would ultimately be good for peasants, who had previously merely 'vegetated' on their farms: 'They were comfortable in their silent vegetation, and but for the Industrial Revolution they never would have emerged from this existence, which, cosily romantic as it was, was nevertheless not worthy of human beings.'[58]

Was Engels right? And even if so, was he not guilty of judging the peasants by his own standards, rather than by theirs? Sadly, we'll never know what the peasants thought, since their views have died with them. The narrative almost always belongs to the powerful, which is what makes the 2012 BBC documentary *The Fastest Changing Place on Earth* so fascinating.[59] Filmed over the course of six years, the film charts the development of White Horse Village, an isolated mountain community 1,000 miles west of Shanghai, into a city of 200,000 residents. Using time-lapse photography, the film shows how the village, with its ancient buildings, muddy lanes and beautiful surroundings, is obliterated to make way for high-rise flats, offices and an imposing party headquarters around a central square.

When interviewed about this transformation, villagers' views are mixed. For one young female rice farmer, the change can't come soon enough: struggling to look after her sick mother and raise two children while her husband works away in Beijing, Xiao Zhang is exhausted. She works night and day, keeping pigs and silkworms in an effort to make ends meet, while yearning for a better life. Six years on, her transformation is remarkable. Now a canteen worker, Xiao Zhang is still exhausted but in no doubt that her new life is far better. For many older farmers, however, the experience is the reverse: many fail to find work in the city and feel useless in their new lives, whiling their time away playing mah-jong in neon-lit cafés while reminiscing about their old life in the countryside. One villager even refuses to sell his house, and sits disconsolately on its bare earthen floor as traffic slams past his door.

China's transformation, like that of Britain two centuries ago, has clearly created winners and losers. While some have made millions from booming construction and manufacturing industries, others have lost their livelihoods and communities without gaining the employment that was supposed to compensate them for their loss. In 2015, just three years after the White Horse Villagers first entered their spartan, jerry-built apartments, the Chinese boom hit the inevitable buffers. Ex-farmers who had blown their compensation money on flat-screen TVs and washing machines couldn't afford the electricity to run them.[60] Some spoke of a sense of *biesi*, feeling stifled to death, in their concrete high-rises; some took their own lives.

Any great upheaval will necessarily inflict pain. Yet however backward (to use the official government term) life for the White Horse Villagers might have been, it clearly had a meaning for some of them that urban life – at least in China's generic instant cities – can't replace.

Will the trauma have been worth it in the long run? Yes and no is the probable answer. In China, as elsewhere, the march of progress usually consists of swapping an arduous life toiling in fields for a similarly arduous existence working in a factory, trading a tight-knit if remote community for the isolation of a flat in a dormitory district. Whether the former is less 'worthy of human beings' than the latter is moot, and since knowing the smell of earth after rain and the names of birds and trees are 'goods' of a totally different order to those of owning a flat-screen TV or Nike trainers, the jury is likely to be out for some while yet.

Capital Gains

As the story of White Horse Village suggests, the claim that urban life is inherently better than rural is too simplistic. Cities clearly offer benefits that the countryside struggles to match: connectivity, opportunity, employment, mobility, variety, culture and access (at least in theory) to markets, schools and hospitals. Cities are cradles of civilisation, yet as places to live they are far from perfect. Apart from the grime, congestion, crime and insecurity, especially if one lives in a slum, there is what economists call the opportunity cost of *not* living in the countryside. Whatever its drawbacks, rural life offers something that most cities can't: closeness to nature.

Contact with nature may seem rather low on our list of priorities as we attempt to scramble up Maslow's hierarchy, but recent research has shown how badly we suffer when we are cut off from it. We recover from illness much faster, for example, when surrounded by greenery, and even short periods of contact with nature have been shown to reduce stress and to increase overall levels of wellbeing.[61] Such findings are hardly surprising – we are after all political *animals*, not political robots, and our dualistic needs go to the core of the urban paradox. Why then do we sign up so willingly to the idea that urban life is better than rural? The answer is that cities offer us hope of economic growth, the very promise that since Adam Smith's day has driven urbanisation.

Rural life is undervalued partly because it neither offers nor relies upon such growth. Indeed, the natural state of rural communities is steadiness. Agricultural yields may rise over time, but will never generate the heady profits to be made from, say, drilling oil out of the ground or clearing rainforest for cattle, an entirely city-led operation. As the Chinese government realised decades ago, the fastest way of generating growth is therefore to turn producers (farmers) into consumers. The White Horse Villagers' blank TV screens and idle washing machines are both the product of such growth and the means by which the next phase will be fuelled. Urbanisation is often credited with lifting people out of poverty, but the reality is often that it simply swaps one kind for another.

The United States is a case in point. At the start of the twentieth century, the US was still a place that John Locke might have recognised: 38 per cent of the population were farmers, and small-town America thrived.[62] Today, less than 2 per cent of Americans live on a farm, and the US has the most industrialised, consolidated food system on earth. The results have been the highest levels of poverty in the developed world and a depressed, drug-dependent, obese population. Those living in rural areas, who represent the fervent core of Donald Trump's base support, are so despairing that suicide rates are three times the national average.[63]

The death of rural America goes to back to Richard Nixon's secretary of agriculture, Earl Butz, who in the 1970s declared war on family farms, vowing to replace them with more efficient agribusiness. 'Get big or get out' was his favoured mantra, along with the equally empathetic 'Adapt or die.'[64] As Michael Pollan explains in *The Omnivore's Dilemma*, Butz backed up his policy by 'subsidizing every bushel of corn a farmer could grow'.[65] The result was a race to the bottom in which farm prices plummeted across the globe.[66] American family farms – once the mainstay of the nation's economy as well as its sense of self – were transformed into a global business in which only the very biggest could survive.

As Joel Dyer describes in *Harvest of Rage*, the destruction of rural America caused widespread trauma. Losing a farm, says Dyer, causes 'incomprehensible despair' in people whose families have worked the same land for generations. Such farmers often feel a responsibility to keep the farm going whatever the cost, and its loss can be more devastating than the death of a loved one. 'You don't just lose a farm,' Dyer remarks; 'you lose your identity, your history, and, in many ways, your life.'[67] The devastation is compounded by the fact that many farmers are traditional, hard-working people for whom continuity and duty matter most.[68] Such people, Dyer notes, 'would often rather die than lose their land'.[69]

Going in Circles

Not everyone fits the novelty-seeking, risk-taking mould of Economic Man, ready to thrive in the entrepreneurial maelstrom of the city. Many farmers occupy the opposite end of the spectrum: for them farming is a way of life that, provided they can earn a decent living at it, they would prefer to keep. Urbanisation is thus far from a one-size-fits-all path to happiness, yet since no alternative to progress is available, it is one that many are forced to take. As a result, villages all over the world are being hollowed out as rural youngsters seek their fortunes in the city, leaving only the very young, sick and elderly behind.

If our narrative doesn't shift, a deep urban–rural imbalance seems inevitable. Yet there could be a better way. As Doug Saunders points out in *Arrival City*, it doesn't take much to transform rural life into something far better. Although Saunders is no urban refusenik – on the contrary, he sees humanity ending this century 'as a wholly urban species' – he also sees the value, at least during the process of transition, of maintaining strong urban–rural relationships.[70]

Urban migration, Saunders points out, is rarely a one-way process, since migrants usually maintain strong links with their home villages, returning to help with the harvest and so on. Most also send cash home to their families, boosting the rural economy. Nowhere has this practice been more pronounced than in Bangladesh, where villages have been transmogrified by money from early migrants to Britain in the 1960s, known locally as *Londoni*. Some villages have grown into virtual towns, replete with shopping malls, cinemas, restaurants, schools and even estate agents.[71] As Saunders explains, *Londoni* are somewhat akin to feudal lords in such villages, living in vast ornate houses and employing up to a hundred people on their farms, in construction projects, shops or road-building schemes.[72] When the river of cash dried up after the 2008 crash, many villages tightened their belts, adjusting from dependent, pseudo-feudal economies into more resilient, self-sustaining ones.

Such village urbanisation, which is already widespread in the wealthier parts of sub-Saharan Africa and the Middle East, could point the way towards a new form of rural economy in the developing world. Rural transformation on the back of urbanisation can give people the chance, should they choose it, of staying on or returning to the land. Such an approach also provides an alternative to the inexorable march of Big Ag. Numerous studies have shown that medium-sized farms in poorer countries are more productive and profitable than larger ones, as well as providing far greater employment.[73] Circular migration not only channels urban wealth back to the countryside, but allows urban migrants to keep a foothold in the country, providing a buffer against the precariousness of their new lives. After the 2008 crash, for example, 20 million Chinese migrants went back to their villages, with 95 per cent returning to the cities later that year.[74] Partly in response, the Chinese government announced in November 2008 that its Ministry of Construction would henceforth be known as the Ministry for Housing and Urban-Rural Planning.[75]

Triumph of the City?

Whatever one's view of urbanisation, it is seemingly inexorable. As cities mushroom and those living in them demand ever more food, farmers all over the world are being turned off the land to make way for the agribusiness that most assume is necessary to feed the urban age. The conflation of progress and urbanisation is thus becoming a self-fulfilling prophecy.

With 7.5 billion people living on the planet and 2.5 billion more expected soon, our options for living 'separated and scattered' are getting ever more remote. Is it time, therefore, to consign the idea of rural life to history? For the US ecologist Stewart Brand, the answer is an unequivocal yes. In 2009, the author of the 1960s *Whole Earth Catalogue* (the green bible of its day) shocked his disciples with the publication of *Whole Earth Discipline*, in which he shatters an entire coconut shy of green shibboleths, arguing that GM crops, nuclear energy and above all urbanisation are our only hope of tackling climate change. Cities are by far the greenest way to live, says Brand, because their density allows 'economies of agglomeration', vastly reducing the resources needed to supply people with everyday services. The stark mix of rich neighbourhoods and slums in cities like Rio de Janeiro might seem jarring, but in fact the arrangement is hugely efficient, since all the 'maids, nannies, gardeners and security guards walk to work'.[76] Slums are grim, yet they are also hotbeds of wealth creation and invention, representing an informal economy potentially worth trillions.[77]

Crucially, Brand argues, urbanisation has the great benefit of emancipating women from their traditional roles as 'fetchers of water and fuel' under the sway of their husbands.[78] Financial lenders have long known that women make far better household managers than men: free to earn money and own their own homes, women could revolutionise the oppressive traditions of patriarchy. Such a transition would have the vital benefit of bringing down birth rates, a sure win–win as far as the planet is concerned. Although there is much work to be done converting today's slums into the cities of the future, concludes Brand, the general direction of travel is wholly positive.

Has Brand got it right? Certainly he is spot on about the invaluable benefits of emancipating women. Urbanisation and education are key to the relative gender equality and lower birth rates that rank among the West's greatest social achievements. Yet does such emancipation necessarily depend on cities? In Kenya, for example, Kikuyu women migrate to the city because they have no property rights in their traditional communities. However, as the social scientist Diana Lee-Smith found, many would prefer to stay in their villages if their rights there were assured.[79] Urban life, for such women, is Hobson's choice – an imperfect solution to complex social questions. A better response might be to persuade tribal elders to alter traditional property laws, an effort that is already under way. Such a solution would offer the women a real choice about where they want to live, rather than impose one on them in the name of progress.

For Brand, however, such choices are 'unaffordable'. He believes, like Dickson Despommier, that the ecological challenges we face mean that withdrawal from nature is our only option. Both men have a point: we have consistently proved ourselves to be a uniquely destructive (and self-destructive) species. But does it necessarily follow that our best chance of survival, never mind of leading good lives, is to move into cities and pull up the drawbridge? Might not this risk alienating us from the natural world upon which we all depend?

Such questions bring us back to the core dilemma of how to feed ourselves in the future. For Brand, the answer is clear: we must farm as intensively as possible and use genetic engineering – GE, a term Brand prefers to GM, since he argues all evolution has involved the latter – in order to minimise our agricultural footprint and release as much of nature as possible back to the wild.[80] Subsistence farming, for Brand, is nothing more than 'a poverty trap and an environmental disaster'.[81]

Although subsistence farming can indeed be a poverty trap, Brand's solutions raise more issues than they resolve. As Raj Patel points out in *Stuffed and Starved*, the blind application of new farming technologies can cause far greater harm than good. The so-called Green Revolution, for example, in which India's Punjab was transformed by a US-backed initiative to improve crop yields through the use of high-yielding varieties, irrigation and chemical fertilisers, led to spectacular results at first, but then went disastrously awry as aquifers ran dry and soils became salinised. Thousands of farmers, unable to pay back their loans, committed suicide. The Green Revolution failed, Patel notes, because it relied on technologies evolved in faraway labs that took no account of local conditions.[82] In contrast, a 1950s programme of rural development in Kerala that involved incremental land reform, the creation of farmer collectives and a new state health and education programme led to both sustained higher crop yields and lasting improvements across the whole of society.[83]

As with any ideology, the problem with urban evangelism lies in what it edits out of the frame. The view that most subsistence farmers would be better off doing something else may well be true, yet the conclusion that they should all stop farming doesn't stand up to scrutiny.

Such people are among the most important on earth, since their knowledge and skills in nurturing nature are vital to all our futures. We only care about what we know, so if we all move into cities, where will such knowledge come from? Nobody has a greater understanding of the land or more incentive to preserve it than families who have farmed it for generations. If conserving nature is our goal, then trusting Big Ag seems a strange way to go – a bit like asking wolves to look after your favourite sheep.

What the do-or-die urban vision ignores is the potential to create new kinds of green communities, both in the city and in the country. It also overlooks the fact that, should today's three billion rural dwellers all urbanise and adopt Western lifestyles, it would spell ecological catastrophe. Cities may be theoretically green, but once their inhabitants start guzzling burgers, driving SUVs and upgrading their phones every year, the ecological gains of everyone living cheek by jowl would rapidly vanish. None of this is to say that the world's poor, two billion of whom already live in cities, don't deserve just as good a life as the rest of us. On the contrary, our most urgent task now is to come up with a vision of such a life that we can *all* enjoy without trashing the planet.

Et in Arcadia Ego

> Human society and the beauty of nature are meant to be enjoyed together.
>
> Ebenezer Howard[84]

Our future is urban, but what does that actually mean? Will cities in future resemble those of the past, or morph into something else entirely? Our current urban models evolved, after all, when the vast majority lived in the countryside. Given our renewed need to live within our ecological means, piling into ever-vaster consumption cities on the model of ancient Rome seems an obvious blunder: plundering distant lands is how we got into this mess in the first place. If we're looking to the past for inspiration, the city-state seems to hold far more promise, not least because it has been the consistent choice of those who have worried about uncontrolled urban growth or running out of land.

Concerns about self-sufficiency made the ancient Greeks limit the size of their *polis*; after Rome's fall, medieval Italians built cities that were small but perfectly formed. London's vastness would later inspire two key utopian tracts: Thomas More's 1516 *Utopia*, a satirical sideswipe at the Tudor capital's greed, and Ebenezer Howard's 1902 *Garden Cities of To-morrow*, a response to the overcrowding and poverty of the Victorian metropolis. Both utopians saw London's dominance as an existential threat, and both proposed a return to the city-state – with an ideal population of around 30,000 – as the solution.

While clearly eager students of Plato and Aristotle, More and Howard disagreed with their ancient mentors on the question of who should farm. While the Greek philosophers had assumed that this would be done by slaves, the English utopians were at pains to present farming as a pursuit worthy of anyone. More's Utopians spent at least two years farming, with many enjoying it so much that they chose it as their life's vocation. While Howard was less enamoured of farming *per se*, probably due to a disastrous stint trying it out for himself in Nebraska, he nevertheless presumed that his farmers, who were to number 2,000 per garden city, would lead just as good lives as their urban counterparts.

Of all utopians, Howard arguably did more than any to address the urban paradox head on. After his Nebraskan fiasco, he spent four years in Chicago, experiencing the agri-industrial boom for himself. Returning to London in 1876, he found his countrymen in the grip of a deep depression caused by cheap imports from the very place he had just left. Shocked by the poverty and crowded slums, he resolved to find some way of steadying the rural economy. The solution, he thought, was to create a network of new urban centres in the countryside where land could be easily bought due to the depression. Connected by railways, such 'Town–Country Magnets' (soon rebranded with the rather catchier 'Garden Cities') would provide enough scale and density to support civilised life, while allowing people access to nature, thus giving them the best of both worlds:

There are in reality not only, as is so constantly assumed, two alternatives – town and country life – but a third alternative, in which all the advantages of the most energetic and active town life, with all the beauty and delight of the country, may be secured in perfect combination.[85]

While wondering how to fund his vision, Howard stumbled across the work of the American economist Henry George. Having witnessed the coming of the railroads in the US and the parallel rise of the 'robber barons', George had worked out that it wasn't the building of railways that made the barons rich, but the vast rise in land values that occurred wherever the railroads went, earning vast fortunes for landowners at no cost to themselves.[86] In his 1879 book *Progress and Poverty*, George argued that the reason economic growth was always accompanied by increased levels of inequality was because the wealth it generated went not into higher wages but into land:

In the great cities, where land is so valuable that it is measured by the foot, you will find the extremes of poverty and luxury. And this dis- parity in condition between the two extremes of the social scale may always be measured by the price of land.[87]

To Howard, George's ideas were a revelation. If the land on which his garden cities were built was owned by a community trust, then any increase in land values would go, not to private landowners, but back to the city. Over time, the community would not only be able to pay off its debts and become economically independent, but would have enough coming in from residents' rates to run services such as health and education.[88] Crucially, such an arrangement would also prevent the city from encroaching on its own farmland, a 'disastrous result' that would be inevitable if the land were 'owned by private individuals anx- ious to make a profit out of it'.[89] While the city would need to import some of its food, its farmers would still make a decent living, due to their advantage of proximity to the market (Howard, to be fair, had never encountered Walmart or Tesco). Run by a board of management to be chosen from within the community, the garden city would effec- tively become a semi-independent city-state.

Far from the bucolic, sleepy image it conjures up today, the garden city was a radical proposal to subvert the capitalist order through incremental land reform and common ownership – the clue lies in Howard's original 1898 title *To-Morrow: A Peaceful Path to Real Reform*. But perhaps the most remarkable thing about garden cities was that, unlike most utopian visions, they were in serious danger of getting built.

Lessons from Letchworth

By 1903, Howard's Garden City Association boasted 1,300 members, including leading industrialists such as George Cadbury, Joseph Rowntree and W. H. Lever.[90] The following year, the committee bought a 3,818-acre plot near Letchworth, a Hertfordshire village just 34 miles from London, employing renowned architects Parker and Unwin to draw up a masterplan. It was a triumphant moment for Howard, one that would, however, prove the zenith of his dreams.

Flaws soon began to materialise. Howard had assumed, mistakenly, that investors would flock to his project, despite promised returns (capped at 4 per cent) well below the usual rate. Howard also infuriated the board by refusing to compromise, prompting a satirical letter from his friend George Bernard Shaw, comparing him to the failed utopian Robert Owen:

> If, like Owen, you insist on wasting on amateur socialism the few thousand pounds which you will no doubt get as easily as he did from sentimental millionaires, then you will fail as he and many others failed. They said all that you say; they knew all that you knew; they had much more experience of manufacturing business; they were just as clever, just as eloquent, just as high-minded, just as right on paper as you are; but they failed, because they did not see that since social-ism is the alternative to private enterprise, a private enterprise which aims at socialism is simply aiming at suicide.[91]

With typical wit, Shaw had nailed Howard's problem: for its backers, the garden city was little more than a feel-good enterprise whose social vision should never be allowed to harm its bottom line. Howard was sidelined, writing indignant letters to the board while his cherished principles – community rents, communal ownership and true self-government – were jettisoned one by one.

By the time Letchworth was built, Howard had largely washed his hands of it. Although the garden city was not without its successes – the world's first green belt and a community trust which, although never quite matching the heights of medieval Siena, operates to this day – it fell far short of Howard's original ambition. His dream of turning England's green and pleasant land into a hotbed of self-governing city-states instead produced a genteel, leafy, arts and crafts town whose middle-class residents mostly commuted to London.[92]

The garden city failed, not because it was a bad idea, but because it made false assumptions about what people really wanted. More than feeling good about themselves, Howard's investors wanted to make money; rather than run their own city-state, people just wanted a nice place to live. With the latter, however, Howard really did succeed: Letchworth remains hugely popular among its residents.

As Letchworth suggests, building communities from scratch is fiendishly hard. The path to utopia is strewn with such misfires; indeed, the word itself (meaning 'good place' or 'no place') bears the etymological seeds of its own destruction.[93] If we're going to live differently in future, we're going to need a more integrated approach than Howard was able to muster: one that corrals more than a few willing industrialists, and can harness serious flows of money, energy, food, employment, transport and services. In short, we're going to need *planning* – which, by virtue of being an inherently socialist activity, has been notably absent in the neoliberal world in recent decades.

Geography of Capital

London's recent property boom is a case in point. As Anna Minton argues in *Big Capital*, London's recent transformation has been caused not so much by a sudden surge of people wanting to live here, but rather by a global spending spree among those looking to deposit cash in London's highly deregulated housing stock. In 2017, the group Transparency International reported that a staggering 44,022 London properties were owned by overseas companies, 90 per cent of which had been bought via secrecy jurisdictions such as the British Virgin Islands and 986 of which were linked to 'politically exposed persons'.[94] London, in short, is the money-laundering capital of the world, its poshest addresses little more than brick-clad safety deposit boxes for kleptocrats and oligarchs looking to launder the proceeds of their illicit wealth.

All the cash pouring into London has created a ripple effect, pushing ordinary Londoners out of the city, leaving the centre as a playground for the rich. It's not just foreign investors getting in on the act either; councils have been compulsorily purchasing their own housing estates in order to get some much-needed cash. Residents from the recently demolished Heygate Estate in Southwark have ended up as far afield as Slough and Rochester, some twenty miles away from their original homes.[95] The exchange value of London residential property now so far exceeds its use value that most new properties are snapped up by foreign investors before they are even built. As Minton notes, 'the UK housing market doesn't function like a pure market; it is linked to global capital flows, not local circumstances'.[96]

Just as cities such as London have become luxury playgrounds, swathes of countryside are going the same way. In 2012, a study by Dutch architect Rem Koolhaas of a strip of countryside just north of Amsterdam found that, of fifty-seven businesses in the area, just eleven were farms, including a 'Cow Hotel', in which 150 dairy cows slept on water beds and were fed by robots.[97] Among the non-farm businesses were a sculpture garden, heritage mill, estate agent, tax consultant and relaxation centre. The study included the hamlet of De Rijp, a picture-postcard fishing village identified as a heritage site by UNESCO, but where you'd struggle to find any fish. 'Like a mini-Venice', all De Rijp's former shops and services have been stripped out and replaced with businesses suited to its new role as a rural relic: a tourist office, herring museum, café and art galleries. Although the local estate agent had farms for sale, these were residences at exorbitant prices for 'Amsterdammers who want a farm but don't actually want to *farm*'.[98] The irony, said Koolhaas, was that wealthy urbanites were moving to the area 'attracted by its aura of authenticity', seemingly oblivious to the fact that the very quality that drew them was being eroded by their own 'urbanising presence'.[99]

Pastoral fantasy is nothing new; it is arguably as old as cities them-selves. Yet, as Koolhaas points out, our fake vision of the countryside masks the stark reality of the vast enterprises that feed us, operating from buildings more rigid, ruthless and factory-like than anything to be found in cities. Far from being the rural idyll of our imagination, the countryside is increasingly a dumping ground for urban externalities, strewn with delivery centres, server farms and power stations like so many toys thrown out of a giant pram. The modern countryside, as Koolhaas notes, is a *terra incognita*: a place about which we know very little and about which our ideas are both 'utterly unreliable and utterly manipulated'.[100]

Otium and *Negotium*

Reality in the modern world is often not what it seems. The old dis-tinction between city and country has largely disappeared, replaced by a new divide between luxury and poverty. Wealth buys you access to both 'city' and 'country', yet neither is real, since the respective essences of both realms – public life and agriculture – are largely absent. The rich can nevertheless indulge in fantasy versions of urbanity and rural-ity, while the poor are banished to spaces beyond the stage set, where the real machinations of society play out. Presumably those who pre-sided over London's property boom like their cities bereft of corner shops and lined with Italian marble, but if we want to create a more grounded, heterogeneous society, the free market clearly won't get us there.

Planning, as Lorenzetti's *Allegory* suggests, is all about land and power. Deciding who gets to do what where goes to the core of politics, since in order to flourish, we all need space in which to live, work and play. Planning is thus political philosophy made spatial. Ownership is clearly central to it, something that Howard tried, yet ultimately failed, to address with his garden city. But planning also requires vision – an idea of what an ideal landscape might actually look like.

One thing we know is that, in the sort of zero-carbon steady-state economy we need to aim for, the land will look very different to the way it does now. To live within our ecological means, we shall need to build more localised, resilient communities based on the principles of *oikonomia*. Since economic growth will no longer be the driving force in our lives, we'll need landscapes that let us grow in other ways, through the increasing richness and agency of what we do. Since land and labour will remain our only sources of wealth, we'll need to make effective use of both. In short, we need to reimagine the landscape as a canvas for human flourishing.

On the question of how such a landscape might organise itself spatially, utopians have been pretty much unanimous: city and country should be as closely linked as possible. This is not just for economic reasons, but because as political animals we need both society and nature. For as long as cities have existed, those who could afford it have always kept a place in both city and country. Wealthy Romans, for example, regularly left the capital for their rural villas, combining *nego-tium*, business, with *otium*, rest. As well as serving as contemplative retreats, such villas were highly productive: 'I cultivate my fruit trees and fields as carefully as my vineyards,' wrote the younger Pliny, 'and in the fields I sow barley, beans and other legumes.'[101] The Roman elite, in other words, practised a luxurious form of *oikonomia*, as the landed gentry of every nation has done ever since. For the majority limited to just one home, however, the struggle is on to find somewhere to live that has access to both social and natural resources.

One reason the city-state model has proved so enduring is because that is precisely what it offers. We Brits may love our suburbs with their semis and back gardens, yet few utopians have approved of such urban sprawl. Suburbia doesn't solve the urban paradox because it doesn't deliver urbanity or rurality at a sufficient scale or density: instead of being the best of both worlds, it offers only penumbra. As utopians from Aristotle to Howard have recognised, in order to be truly urban, cities need to be a certain size and mass.[102] Although what constitutes 'big' or 'small' today has changed somewhat (one would need to multiply the Greek ideal by at least ten), the imperative for high-density cities remains the same, since they deliver urban life without taking up too much land. Since we can build upwards as well as sideways, balancing city and country is as much a question of pattern as of scale: Manhattan, for example, delivers a huge dose of urbanity in a remarkably small space, with Central Park to provide some green relief.

Although Central Park isn't a farm, like all other parks it has the potential to become at least partially productive. Whether or not they contain fruit or nut trees, such spaces represent a principle common to all great cities: the need to balance density with open space. Whatever form such spaces take, whether they are rivers, gardens, squares or courtyards, they are vital to a city's liveability. London's green belt, established by Patrick Abercrombie in his 1944 London Plan and arguably the garden city's most famous descendant, is one example of this principle. Tokyo, meanwhile, has its own urban–rural mix in the form of a patchwork of organic farms, which, thanks to Japan's similarly far-sighted 1952 Agricultural Land Act, are unlikely survivors in the heart of the megacity.[103] Today, however, most urban development across the world is unplanned, driven by the same capitalist forces that did for Letchworth.

The Neotechnic Habitat

> We must . . . make the field gain on the street, not merely the street gain on the field.
>
> Patrick Geddes[104]

Future urban growth is inevitable, yet the shape of it is not. If our goal is to create a world in which everyone can flourish, then our common task must be to plan what form such growth might take. Given that we already have a vast library of historical examples at our disposal, we at least have a chance of getting this next Great Transformation right. As Kate Raworth argues in *Doughnut Economics*, our approach to any future development will need to be two-tiered, since those in the Global South need to increase their living standards and thus their global footprint, while we in the North need to reduce and stabilise ours.[105] In creating a resilient landscape, therefore, our task is similarly split: while in the South it largely involves managing explosive urban growth, in the North it is more a case of post-fitting the cities and countryside we already have. In either case, creating landscapes for future human flourishing will depend on finding ways of maximising the urban–rural interface.

This idea isn't new; it was first proposed more than a century ago by the Scottish geographer, biologist and father of regional planning, Patrick Geddes. Human development for Geddes was all about the natural landscape, the 'active, experienced environment' that shapes everything we do.[106] Like Howard, he deplored urban sprawl, in particular the tendency of cities to merge into one continuous metropolitan splurge, a phenomenon for which he coined the term 'conurbation'. His solution was to protect rural areas close to cities, not by a green belt but with a series of rural 'fingers' radiating out from the centre, making the city form a star shape as it expanded. By this means, he hoped to create cities truly in touch with nature: 'Towns must now cease to spread like expanding ink-stains and grease-spots: once in true development, they will repeat the star-like opening of the flower, with green leaves set in alternation with its golden rays.'[107]

Despite his dislike of sprawl, Geddes was no technophobe; on the contrary, he imagined a 'Neotechnic Age' on the horizon in which new technologies such as electricity, cars and telephones would liberate people from the shackles of geography, allowing them to live in society as well as close to nature. What the Neotechnic Age has actually brought, of course, is sprawl on a truly epic scale, with conurbations that stretch for hundreds of miles up the US East Coast or across China's Pearl and Yangtze River Deltas. Would Geddes have despaired? Given his Tiggerish outlook, the chances are that, on the contrary, he would probably have said that we've finally got the wherewithal to transcend space; we're just being rather slow at applying it.

Such a view isn't wrong: for the first time in history we do in fact hold the key to the urban paradox. We can live in the middle of nowhere, yet still hold down an office job, stay in touch with friends and family, read the news, visit a library or bank, catch a movie or see a lawyer or doctor. The Internet is a remarkable social tool, yet the irony is that it has destroyed the places and institutions that once brought us together: markets, high streets, local banks and libraries. It has phenomenal power to connect us, yet in doing so distances us from the physical world. This is most obvious when one is out on a beautiful day, as I was recently in a London park, where about half the people I passed were wearing headphones or squinting at their phones, oblivious to the sunshine, blossom and birdsong all around them.

Mobile phones can of course have a hugely positive effect, as for example in Kenya, where farmers can check market prices from their farms and transfer money through e-currencies such as the Kenyan M-Pesa. Such technologies are already transforming lives, yet what few are asking is how they might help us find transformative ways of inhabiting land. Which brings us back to the question that underpins all utopian visions and indeed the very question of how we should live – food.

Lessons of War

> Buy land: they aren't making it any more.
> Mark Twain[108]

By 2050, the global food map is likely to have changed beyond recognition, with water stress, desertification and failed harvests due to extreme weather creating havoc for which few are prepared. Nowhere is less prepared than the UK. Used to importing our food from all over the world, we assume that others will carry on feeding us come what may, yet, as three leading UK food experts warned in a 2017 paper entitled *A Food Brexit: Time to Get Real*, that assumption could prove dangerously wrong. In an overheated, overpopulated, urbanising future, they argued, our traditional suppliers may be too busy feeding themselves to bother with us. Britain, which currently imports one third of its food from the EU, is sleepwalking into a potential crisis. 'The UK food system ought to be improving its resilience,' they wrote. 'It isn't. It's like the rabbit caught in the headlights – with no goals, no leadership, and eviscerated key ministries.'[109]

Instead of assuming we'll always be able to import food from abroad, the authors went on, we should be transitioning towards a more sustainable diet in the UK, which is to say one that is less wasteful, meaty and carbon-producing and more regional and seasonal.[110] Recognising that such proposals were likely to put a political cat among the pigeons

(Brexit, after all, was supposed to herald a glorious free-trade future), the authors nevertheless raised questions that demand answers. Whatever the eventual outcome of Brexit, the question of how many pony parks we might have to shut in order to grow more cabbages could come sooner than we think.

It is no accident that the last time we addressed ourselves to such issues was during the Second World War; it took a crisis of that magnitude to make British politicians take food seriously. In 1939, British farms were producing just one third of the nation's calories and five out of six still relied on horsepower.[111] It is testimony to the power of a crisis that by the end of the war the amount of farmland in Britain had doubled, the number of tractors had quadrupled and food production had risen by one third. Although nobody liked being bossed around by the Ministry of Food, which oversaw every aspect of the national food supply from production and trade through to education and rationing until 1958, the result was that the nation had never been healthier.

In the space of just six years, while attending to the pressing matter of defeating Hitler, Britain transformed itself from an uninformed, poorly fed, heavily dependent food nation into a knowledgeable, healthy, far more resilient one. We managed it, of course, because our lives were on the line, which is in effect where we are again, although today we face a common threat far more lethal than a genocidal fascist. In order to survive, we need to transform ourselves into an engaged, egalitarian force of motivated, self-reliant citizens. Can we do it? Since the British invented the industrial-capitalist model that got us into this fix, one might call it our moral duty at least to try.

Property is Theft

> . . . the fruits of the earth belong equally to us all, and the earth itself to nobody!
>
> Jean-Jacques Rousseau[112]

One advantage we have is that, thanks to our pre-industrial past, we already know how to thrive in a steady-state economy. In essence, we need a modern version of the *polis* minus the patriarchy and slaves. Such a society would be based on *oikonomia* and would be green, resilient, democratic and egalitarian. It's pretty clear what we need; what's much less obvious is how to get there.

For the French philosopher Pierre-Joseph Proudhon, the answer lay in common ownership. In his 1840 essay 'What is Property?' Proudhon argued that rulers and property were antithetical to a good society, since they undermined the essential tenets of equality and freedom. 'If slavery is murder,' said Proudhon, then 'property is theft,' since nobody is born with authority over others, or owning anything. Although we all need access to property (most notably land) in order to live, none of us has the right to call it ours for ever, since others in future will have an equal need. So although we can *possess* land during our lifetime – put, as it were, our symbolic towel on the metaphorical deckchair – we can never own it outright, since it ultimately belongs to society as a whole.

Proudhon's ideas inspired a new political movement, anarchism. Although 'anarchy' today is a word synonymous with chaos, for Proudhon it simply meant a society without rulers (from Greek *a*, without + *arkhi*, ruler), one that was therefore truly cooperative and egalitarian. When Proudhon summoned up his human *tabula rasa*, he saw neither Hobbes's rabble nor Locke's agrarians, but something far closer to our true hunter-gatherer past: small groups of territorial collaborators without formal leaders, a society whose very existence depended on common ownership.

In refutation of Locke, Proudhon asserted, 'Neither labour, nor occupation, nor law, can create property.'[113] Since populations were always in flux, he reasoned, the amount of food, land and other resources available to each person must fluctuate too. 'Would it not be criminal,' he asked, 'were some islanders to repulse, in the name of property, the unfortunate victims of a shipwreck struggling to reach the shore? The very idea of such cruelty sickens the imagination.'[114] Clearly the only civilised response of the islanders would be to welcome the newcomers to live with them, tightening their belts accordingly. With this direct appeal to arguably the deepest expression of human civility – hospitality – Proudhon argued that in order to create a just society, property must be abolished.

Without property, Proudhon asked, what form should society take? Might it be communist? The problem with that, he said, was that although communism *pretends* to abolish property, it does so 'under the direct influence of the proprietary prejudice', with the result that 'life, talent and all the human faculties are the property of the State'.[115] Far from delivering freedom, said Proudhon, communism merely led to 'oppression and slavery'. Indeed, 'the disadvantages of communism are so obvious that its critics never have needed to imply much eloquence to thoroughly disgust men with it'.[116]

Neither communism nor capitalism could deliver a good society, said Proudhon, since the former rejected 'independence and proportionality', while the latter negated 'equality and law'.[117] What was needed was a political third way that combined the benefits of both: 'This third form of society, the synthesis of communism and property, we call *liberty*'.[118] All that was needed to create such a system, said Proudhon, was to abolish the principle of proprietorship: 'suppress property while maintaining possession and you will revolutionise law, government, economy, and institutions; you will drive evil from the face of the earth'.[119]

Anarchy in the UK

> Well-being for all is not a dream.
>
> Peter Kropotkin[120]

Proudhon's fluidity of thought resonates in our overcrowded world. His shipwrecked survivors are no longer the stuff of metaphor, but real migrants risking death to seek better lives on our shores. In contrast to Locke's picket-fenced rigidity, Proudhon's belief that the more of us there are to share the pie, the smaller each slice must be, is strikingly modern. Locke of course never imagined that we could run out of land, while for Proudhon the possibility was all too clear. His solution – a flexible system of shared ownership – was bold, yet on the matter of how to get there, he was as naïve as Paddington Bear. Once the bourgeoisie realised how iniquitous property ownership was,

Proudhon believed, they would gladly give up their luxury pads for the benefit of all. Needless to say, he was to be sadly disappointed in this hope.

It would fall to Proudhon's follower, the Russian exile Peter Kropotkin, to fully grasp the anarchist nettle. Since all wealth and property were the result of society's former labour, Kropotkin argued, they must belong collectively to everyone. 'There is not even a thought, or an invention,' he wrote in his 1892 book *The Conquest of Bread*, 'which is not common property, born of the past and the present.'[121] The fact that such property had ended up in the hands of just a few was clearly wrong; the people must take back what was rightly theirs. 'There must,' said Kropotkin, 'be *expropriation*.' Amazingly, Kropotkin shared Proudhon's optimism that such a 'revolution' (a term he used openly) would be peaceful, a belief that proved just as unfounded as Proudhon's trust in self-sacrificing aristocrats. Unsurprisingly, the closest thing to a functioning anarchist society yet to emerge – Catalonia 1936–9 – came in response to war.

In his 1938 memoir *Homage to Catalonia*, George Orwell described how in 1936 he had gone to fight in the Spanish Civil War. Arriving in Barcelona, he found the city transformed: since many residents had fled at the outbreak of hostilities, the city was being run by anti-fascist forces and anarchists.

> It was the first time that I had ever been in a town where the working class was in the saddle. Practically every building of any size had been seized by the workers and was draped with red flags and with the red and black flag of the Anarchists . . . Every shop and café had an inscription saying that it had been collectivized; even the bootblacks had been collectivized and their boxes painted red and black. Waiters and shop-walkers looked you in the face and treated you as an equal . . . In outward appearance it was a town in which the wealthy classes had simply ceased to exist . . . Practically everyone wore rough working-class clothes, or blue overalls or some variant of military uniform. All this was queer and moving. There was much in it that I did not understand, in some ways I did not even like it, but I recognised it immediately as a state of affairs worth fighting for.[122]

This state of affairs was, however, to be short-lived. By the time Orwell returned from the front in April 1937, revolutionary fervour had noticeably waned and the city was returning to business as usual. The anarchist society he had witnessed, Orwell realised, had been partly an illusion, a 'mixture of hope and camouflage' in which the terrified bourgeoisie had 'deliberately put on overalls and shouted revolutionary slogans as a way of saving their skins'.[123] In the countryside, however, more genuine reform was taking place, as workers organised themselves into collectives to run farms, shops, factories and businesses. In certain rural villages, money was abolished and replaced by a local points system. Although some no doubt had their misgivings about the new arrangements, the anarchist flame nevertheless briefly flickered.[124]

Anarchism is inherently unstable. Without the usual rulers and hierarchies and reliant instead on mutual cooperation, it is the purest form of democracy, and as a result almost unfeasibly utopian. For Kropotkin, such fluidity was anarchism's great strength, allowing societies to resist their usual tendency to 'ossify into fixed and immovable forms', leaving them free to transform like a 'living organism continually in development'.[125] An anarchist society, he believed, would adapt just like a natural ecosystem.

Kropotkin recognised, as Rousseau had done, that the functioning of true democracy required rounded, capable citizens; creating such people was, for him, society's entire *raison d'être*. Foreshadowing Schumacher, he argued that capitalism's core article of faith, the division of labour, had done immeasurable harm. In order to become complete, fulfilled individuals, Kropotkin argued, people needed to have diverse lives: to work with their hands as well as their heads, both in workshops and on the land. 'Political economy has hitherto insisted chiefly upon division,' he wrote. 'We proclaim *integration*.'[126]

Like Proudhon, Kropotkin foresaw the crisis of globalisation with uncanny precision. Once developing nations industrialised, he reasoned, they would create their own internal markets and no longer be interested in 'supplying us with our staple food and luxuries'.[127] China, in particular, would develop its own market for luxuries. 'China will never be a serious customer to Europe,' he wrote; 'she can produce much cheaper at home; and when she begins to feel a need for goods of European patterns she will produce them herself.'[128] Industrial capitalism, in short, had a beginning, middle and end, and it was therefore essential to look beyond it:

> We have before us a fact of the consecutive development of nations. And instead of decrying or deposing it, it would be much better to see whether the two pioneers of the great industry – Britain and France – cannot take a new initiative and do something new again . . . namely, the utilisation of both the land and the industrial powers of man for securing well-being to the whole nation instead of the few.[129]

In his 1898 book *Fields, Factories and Workshops*, Kropotkin laid out his vision for such integrated productive landscapes: the eponymous fields, factories and workshops that would create a network of self-organising communities. No longer the 'monopoly of the major cities', art, science and industry would be scattered throughout the countryside, much as crafts and trades were in the pre–industrial era.[130] Farming would be central to the new economy; indeed, Kropotkin devoted several chapters to working out how the UK might feed itself, something he thought could easily be achieved. Farming was, however, about far more than feeding the nation; it was also a perfect way to engage with the joys of nature. 'You will be astonished at the facility with which you can bring a rich and varied food out of the soil,' he wrote. 'You will admire the amount of sound knowledge which your children will acquire by your side, the rapid growth of their intelligence, and the facility with which they will grasp the laws of Nature, animate and inanimate.'[131]

If any of this sounds familiar, it is because Kropotkin was the other key influence, along with Henry George, on none other than Ebenezer Howard, as he was dreaming up his garden city. Of all the notions that might spring to mind when strolling through the leafy lanes of Letchworth, anarchism is probably the last, yet it is the foundational idea lurking beneath each neatly thatched roof and behind every trimmed privet hedge.

The Value of Land

> We must make land common property.
> Henry George[132]

Anarchism has been around for nearly two centuries with little to show for itself, so why bother with it now? The answer is that its time has arguably come.[133] As populism rises and capitalism totters, we are in greater need than ever of a social vision that transcends the fatal duality of neoliberalism and totalitarianism, one capable of engaging with and connecting us at every scale from local to global. Although a full-blown anarchist society would be near-impossible to build or sustain, anarchism's core message, that we should embrace democracy while sharing our goods more in common, could hardly be more apposite. By accepting our duty as political animals, it suggests, we can become more effective, empathetic, fulfilled social beings.

As Locke and Smith did before them, the anarchists realised that all human prosperity rests on land. Whether we live on it directly or simply use it as part of our ecological footprint, land (aka nature) is what sustains us, yet the amount that we use differs enormously: if we all lived like Americans, it would take an estimated four planets to keep us going.[134] Ownership is also critical. One third of all land in Britain, for example, is owned by the aristocracy, one of the factors that perpetuates the structural inequalities of our society.[135] If we are all to thrive in the future, it's clear that we shall need a radical reform of the ways that we use, share and inhabit land.

The anarchists' big idea was to abolish private property altogether, an act that strikes at the heart of capitalism and our very idea of a good life. Such a proposal is unlikely to find many takers, yet its principles are vital if we are to avoid social and ecological meltdown during this century. But is it possible to even up the property playing field without resorting to revolution or war? Henry George's *Progress and Poverty* might provide the answer. George, you may recall, worked out that the reason progress was seemingly always accompanied by increasing poverty was that the wealth it generated ended up in the rising value of land rather than in the hands of workers. His solution (which Howard tried, yet failed to fully implement) was to place all land under common ownership and charge individual landowners for the privilege of using it:

> I do not propose either to purchase or to confiscate private property in land. The first would be unjust; the second, needless. Let the individuals who now hold it still retain, if they want to, possession of what they are pleased to call their land. Let them continue to call it their land. Let them buy and sell, and bequeath and devise it. We may safely leave them the shell, if we take the kernel. It is not necessary to confiscate land; it is only necessary to confiscate rent.[136]

With this simple idea George came up with a way of neutralising rising inequality at a single stroke: a land-based wealth tax. By taxing the value of land – effectively charging a community rent for it – society could both share its wealth far more equitably and make better use of the available land. Since the tax on owning land in city centres would be high, its purchase price would be greatly reduced, making it more affordable. City landowners would have an incentive to develop empty plots, thus helping to increase urban density and so prevent sprawl. Farmland would become cheaper too, since its speculative value would be nullified, making it more accessible to would-be farmers. Since no land could be shifted offshore, the ancient art of tax avoidance would be at an end. So confident was George in the success of his idea that he proposed to 'abolish all taxation save that upon land values'.[137]

Progress and Poverty was an instant global sensation, coming second only to the Bible in sales and inspiring the likes of Howard along the way. George had shown how the anarchist vision could be realised without the need for pitchfork-wielding mobs, achieving the double whammy of redistributing wealth while restoring the natural balance of city and country. Land value tax remains a powerful, simple idea, so why aren't we doing it? The answer is the usual one: it would be wonderful for everyone apart from existing landowners. In the UK, where home ownership is the pillar of wealth, land value taxes seem to threaten the very foundation of democracy. In reality, however, all they would do is stop people accumulating far more land than they need. By reducing the effective value of property close, but not down, to zero, Georgism would achieve what Locke set out to do all along: create an egalitarian, land-based society.[138]

Georgism is one of those ideas that can seem too good to be true. The devil would be in the detail of course, not least in working out how to manage the transition from our current system to one of common ownership. Yet a growing number of advocates including the UK Green Party, the US economist Joseph Stiglitz and British journalist George Monbiot believe that land value taxes could really work.[139] As Monbiot argues, the common ownership of land would not only create a fairer society, but would channel precious resources towards public facilities that could be shared by all:

The new approach could start with the idea of private sufficiency and public luxury. There is not enough physical or environmental space for everyone to enjoy private luxury: if everyone in London acquired a tennis court, a swimming pool, a garden and a private art collection, the city would cover England. Private luxury shuts down space, creating deprivation. But magnificent public amenities – wonderful parks and playgrounds, public sports centres and swimming pools, galleries, allotments and public transport networks – create more space for everyone at a fraction of the cost.[140]

Since they re-establish the land as the true source of our common wealth, land value taxes form a natural part of sitopian economics. By increasing the density of cities while protecting farmland, they could unlock the urban paradox, creating conditions under which a food-based steady-state economy might actually thrive.

The New Commons

The greater the number of owners, the less the respect for common property.

Aristotle[141]

A frequent objection to the idea of common ownership is the 'tragedy of the commons', a notion first propounded by William Forster Lloyd in 1832 and popularised by the US ecologist Garrett Hardin in a 1968 article in the American journal *Science*.[142] The theory (which is as gloomy as that of Malthus, which inspired it) is that the shared use of common resources will always fail, since people sharing, say, fishing ponds or common grazing land will always have more to gain from overexploitation than from taking only their fair share.[143] The only solution, Hardin concluded, was for such resources to be controlled by the state. 'If ruin is to be avoided in a crowded world,' he wrote, 'people must be responsive to a coercive force outside their natural psyches, a Leviathan, to use Hobbes' term.'[144]

For the US political economist Elinor Ostrom, who won a 2009 Nobel Prize for her work on traditional resource management, the 'tragedy of the commons' was poppycock. Although overexploitation could occur, said Ostrom, this was invariably in places where there were no agreed rules for how to share. When such rules were established, shared management of resources was not only highly effective, but frequently more so than any other method. Of ninety-one irrigation systems and fisheries she studied, Ostrom found that over 70 per cent of commonly managed ones performed highly, while only 40 per cent of those managed by external agencies did.[145] Drawing on more than forty years of fieldwork, Ostrom drew up a list of principles for what she termed common pool resource (CPR) institutions. These included clearly defined boundaries, self-determination, participatory decision-making, effective monitoring, graduated sanctions, conflict resolution and, in the case of large-scale resources, appropriate oversight from a higher authority. Most vital of all was mutual trust, established through effective communication and reciprocity.

Although scale was crucial to such institutions, Ostrom found, not every aspect had to be small; indeed an essential element of many was their polycentric character, in which many 'nested' interests and enterprises coexisted. What was essential, however, was an approach that was precisely tailored to local conditions and able to adapt over time. Shared ownership, Ostrom found, not only helped to preserve local resources, but stimulated the kind of political engagement and cooperation needed to build equitable, resilient societies. Of the hundreds of common institutions that she and her team studied, the youngest – an irrigation system in Turkey – had been working for at least a century, while the oldest – Alpine grazing in Switzerland – had been going for more than a thousand years.

Like a bird flying along some invisible wire, Ostrom confirmed one by one all the utopians' favourite tropes. Scale really *does* matter, for all the reasons that Plato and Aristotle espoused, namely to establish trust and to manage resources effectively.[146] Common ownership of land is a win-win too, both encouraging good stewardship and helping to build resilient, engaged communities. Like the anarchists, Ostrom rejected both market and state as effective arbiters of resources; the best stewards, she found, were those with a direct stake in their conservation. Above all, societies had to be diverse and adaptable enough to reflect a wide range of potentially competing interests: 'Panaceas,' Ostrom declared, 'are dysfunctional.'[147]

Ostrom's insights help to show how we might all flourish in a steady-state economy. By reminding us of the fundamental bond between society and land, she points the way to a new approach to global governance. Climate change, like war, could galvanise us into action and remind us of our basic commonality. With that common threat in mind, we could create a new stratum of world governance, a polycentric network of nested local groups and international bodies based on the shared management of common resources. Such a movement would necessarily require the global reform of land and fishing rights, so that food would once again become the dominant force shaping our landscapes and cities and binding us together. Of all the actions open to us, restoring food's true value *across continents* is arguably the most potent and far-reaching. Just as the invention of cheap food once annihilated geography, valuing food again will be key to thriving on our hot, hungry planet.

Et in Sitopia Ego

What would our lives be like if we were to value food again? We can easily find out by studying places where food is still valued, which is to say anywhere that traditional food cultures endure. Whether Alpine pastures, Brazilian jungle markets, Cairo souks, Italian olive groves, French vineyards or Tokyo urban farms, what such places embody is food's capacity to bond us to spaces, landscapes and one another. They reveal food's power to shape our lives and give them meaning over time. Those groves, vineyards and markets have all existed for centuries. If proof were needed that valuing food creates resilient cultures, they are it.

One such tradition is that of the Russian dacha. These are small garden plots with simple wooden houses close to the city, to which many urban Russians repair during the summer months and at weekends to grow some vegetables and relax. The tradition began in the early eighteenth century, when Peter the Great granted loyal vassals country estates near St Petersburg (*dacha* means 'something given'), both to thank them for their services and as a way of keeping them close. While most ordinary dachas are a far cry from aristocratic residences, the habit of regularly leaving the city during the summer has become firmly entrenched in Russia – one that during the depredations of the Soviet era became strongly associated with growing food. Today, some 60 million Russians (40 per cent of the population) own a dacha, and weekend traffic jams out of large cities like Moscow and St Petersburg can be horrendous. The May Day holiday in particular has become an annual mass migration, as millions leave the city to plant their crops for the year.[148] Although food is no longer scarce in Russia, many Russians still love to spend their weekends tending their plots, growing fruit and vegetables for their families and making pickles and jam to enjoy through the winter.

Dachas are one example of how valuing food can bring temporal and spatial order to our lives. In many ways, they replicate the rhythm of *otium* and *negotium* once enjoyed by wealthy Romans. Valuing food pulls us closer to nature, providing a counterbalance to the 24/7 freneticism of urban life. If post-industrial societies like the UK and US were to value food again, the most obvious change would be a rural renaissance. With more people living in the countryside and more cash flowing there, services such as post offices, schools, hospitals, shops and transport would once again proliferate, along with better distributive networks such as markets, freight depots, food hubs and abattoirs. The landscape would also change to reflect a return to smaller-scale mixed-use organic farming. New communities would grow up around food, just as they have always done.

Our cities would also be transformed, with revitalised markets and high streets, more independent shops and restaurants, market gardens and allotments, urban farms, community kitchens and neighbourhood composting. Food planners would work to enhance the diversity and density of both city and country and maximise the interface between the two. Rural areas close to cities would be protected, suburban areas strategically densified, and new compact urban hubs along the lines of garden cities – even the odd vertical farm – introduced on their peripheries.

Is such a vision hopelessly utopian? The evidence suggests otherwise; indeed, as we have seen, it is already happening. The food movement is global and spreading fast, making positive social and ecological change wherever it goes. The whole point of sitopia, indeed, is that by valuing food we can all start building a better world, right here and now. Back in 1973, that was precisely what a group of hippy Brooklyn friends did; today, Park Slope Food Coop is one of the oldest and biggest community food networks in the world, with 17,000 members and a 45-year record of long-term contracts with local farmers in New York State.[149] By putting the *oikonomia* back into economics, Park Slope was a pioneer of the emergent sitopian economy.

Far from its affluent foodie image, however, many of the food movement's most inspiring projects in the developed world are at the opposite end of the social spectrum. Will Allen's Milwaukee-based organisation Growing Power, for example, has transformed deprived neighbourhoods through the teaching and support of community composting, aquaponics and urban food growing.[150] Stephen Ritz's educational gardening programme in some of the toughest schools in New York's Bronx, meanwhile, has had transformative and lasting effects on the pupils who have taken part.[151] And Julie Brown's social enterprise, Growing Communities, in the London Borough of Hackney combines an organic box scheme in some of the city's poorest neighbourhoods with volunteer food growing and an educational programme that teaches people how to help build a more sustainable, ethical food system 'one carrot at a time'.[152] Such projects, of which there are many thousands, help people eat better and create precisely the sort of motivated, knowledgeable, capable citizens of which Kropotkin dreamed.

Crucially, food is also back on the agenda of architects and planners, something our ancestors would be astounded to discover had been forgotten. Food planning is one of the fastest-growing fields in urban and regional design, producing schemes such as Dutch architects MVRDV's Almere Oosterwald masterplan, which incorporates a Kropotkian mix of farms, factories and housing in a deliberately fluid design. London-based architects Viljoen and Bohn, meanwhile, propose filling leftover spaces in cities such as car parks and verges with urban farms to create 'continuous productive urban landscapes' (CPULs): green corridors into the countryside that echo Geddes' star-shaped vision.[153] Cities worldwide are joining forces to move towards more resilient and ethical food systems: the UK's Sustainable Food Cities is one such network, and food also plays a major role in the Transition Movement and the C40 Cities programme, a network of ninety-four leading global cities working together to fulfil the UN's Sustainable Development Goals and combat climate change.[154]

We live in perplexing, perilous, exhilarating times that call for bold ideas and steady heads. Extraordinary new technologies will undoubtedly bring us remarkable capacities in the future, but without similarly daring and innovative social, economic and spatial evolution, they will be worse than useless. It is in this context that food has so much to offer. No matter how thrilling and distracting our digital lives become, food can keep us grounded, reminding us that our fate will always depend on nature, and on how we share it. The way we eat in the future will not only shape our fate, but that of every other species. By valuing food, we can rebalance our lives within the natural world and build happy and fulfilling lives together. After five millennia, we can finally learn to love the urban paradox.

6

Nature

Otter Estuary, Devon

I am staring at some greenish-yellow seed pods in a hedgerow behind one of the most beautiful beaches in Britain. It's a glorious midsummer day, and just beyond my gaze is the sort of scene to make a clammy office worker weep: a sparkling sea, rust-red cliffs and a tall stand of green pines against a cloudless cornflower-blue sky. Yet it is these seed pods – which belong, I have just learned, to the Alexanders plant – on which my whole attention is now focused. My guide is Robin Harford, a renowned forager who sometimes styles himself an ethno-botanist if he can face the barrage of questions that inevitably follows. Today, he is sharing with me some of the secrets of his favourite foraging spot, the Otter Estuary near Budleigh Salterton in Devon.

Robin breaks off one of the pods and hands it to me to taste. There is an immediate explosion of flavour, an intense mixture of celery and black pepper. This, he explains, is because Alexanders belongs to the carrot family, the 'hidden spice rack of the hedgerow'. Unlike commonly cultivated plant families such as mustard, mallow and mint, which are all edible yet mild (edible, Robin observes, 'doesn't necessarily mean tasty'), the carrot clan, which includes parsnips, fennel and celery, is simultaneously 'poisonous and gourmet'. 'Carrot is the real dodgy family,' says Robin, 'but it has all the gold in it, as far as I'm concerned.'

There are at least 700 wild edibles in Britain, but almost all go under the radar. The reasons for this are partly due to the fact that every culture has its own set of approved foods. 'When we became static farmers,' says Robin, 'we didn't have GM labs, so the food plants we started growing came from wild variants. Over ten thousand years we've hybridised and found the plants that have the best flavours; but who's made that decision? The producer, the market, or someone saying *we're going to feed you this*? It's an odd question to ponder.'

Odder still is the fact that many of the plants we ignore have highly desirable properties, such as the Duke of Argyll's tea plant, whose tiny purple flowers are beautiful when viewed through a magnifying glass: five-petalled stars with lime-green centres streaked with purplish crimson. The tea plant is a boxthorn, Robin tells me, a member of the Solanaceae family, to which the fabled goji berry also belongs. Native to Asia, goji berries have long been highly prized in both traditional cookery and medicine, yet have only recently been 'discovered' in the West, to where they are now shipped at vast expense as a 'superfood'. 'Why bring the berries all that way, when we've got a virtually identical plant on our doorstep?' wonders Robin with a note of exasperation. 'Nobody knows whether these berries are as potent as the goji or not, but that's because nobody has bothered to find out.'

Moving a little further up the estuary, we come to the main crop of the day, sea vegetables. The low riverbanks are covered in a dense pale green plant that Robin identifies as sea purslane. We are walking, I realise, through a sea of salad. He bends down and picks a leaf for me to try: the taste is subtle, with that pleasing mix of salt and iron-greenness that all sea vegetables seem to possess. Robin hands me another plant to sample, the first today that I think I recognise, yet out here among all these unfamiliar tastes I don't quite trust my instincts. 'It's samphire, isn't it?' I ask Robin, receiving an affirmative nod. I bite into the stem and receive the familiar hit of sweet green juiciness and maritime minerality. It's delicious: I can see why it has become the go-to vegetable for fish chefs. On this golden day by the seaside, the taste is crazily intense, like a sea symphony dancing in my head. This sensation, Robin assures me, is perfectly normal when eating wild: our senses can go into overload. I am having a ludicrously good time, I realise, with one of those rare, glad-to-be-alive surges of euphoria.

Robin is keen to get to his favourite patch, where sea vegetables grow in sufficient profusion to satisfy even a recent convert like me. We eat sea blite, whose off-putting name belies its delicate fronds and succulent, umami flavours; sea plantain, which, as its name suggests, takes sea-sweetness to new levels; and, last but not least, sea aster, whose elegant leaves and subtle richness make it a prince among sea vegetables. Robin informs me that it is now on sale in Waitrose at £22 per kilo. '*Whaat?*' I say. 'But where on earth are they getting enough of it to supply a supermarket?' I have visions of a coastline denuded of my new favourite vegetable. 'There's a supplier who's cultivating it,' says Robin, 'which I don't have a problem with, but I know plenty of foragers who do.'

The question brings up an obvious issue when it comes to foraging: scale. If we all started harvesting sea aster and picking wild mushrooms tomorrow, the land would soon be stripped bare, which is why most foragers follow a strict code that prescribes when, and how much, they harvest and why 'cultivated wild edibles' may not be such a bad idea. The problem, as Robin points out, is that cultivated wild plants are an oxymoron. 'My only issue,' he says, 'is that, when we take a plant from the wild, it's growing in its optimum conditions, so it's where it's meant to be. Farmers take plants and put them in places, usually, where they're *not* meant to be, so you get pests, you get lower production, smaller harvests possibly; unless you spray it and use all the other funky stuff. Then you have to consider the plants themselves. A wild plant is generally twice as nutrient dense as a farmed plant, so when you take it from the wild and farm it, what are you losing?'

Foraging, it's becoming clear to me, is more than just a way of getting some free food; it's a state of mind. We humans are natural gatherers, Robin explains: our ancestors foraged all year round, knowing that, once they had harvested one crop, others would come into season. Following nature's fruitfulness created an 'abundance mindset' in them, while farming, which often relies on one crop that might easily fail, engenders a fear of scarcity. 'In the past, I would have created twenty jars of hedgerow jam, but as I've developed a deeper relationship with wild plants, I don't need to do that. Now, I don't recommend that people make more than three jars of anything, because as soon as that plant's gone over, there's another twenty wild edibles you can go and work with, so there's no need to hoard.' He reflects for a moment, and then adds with a grin, 'Wild garlic might be the exception.'

Foraging, Robin explains, is 'about feeding and being fed on multiple levels: psychologically, emotionally, mentally, spiritually and definitely physically. You want to get out there as regularly as possible and engage with the plant, so your connection deepens and your wellbeing is enhanced.' It helps, says Robin, that wild herbs and spices don't last as long as shop-bought ones, so the plants 'pull us back into the hedge'. Time is critical to foraging. 'You have short windows of opportunity, so nature teaches us to be opportunists. She gives us the gifts, and if we don't have the eyes to see it and we don't act in that moment, it's over.'

Eating wild, I reflect, is about as far from shopping at Tesco as it's possible to get. While supermarkets are all about blurring time and place – the year-round availability of kumquats – foraging is all about the here and now: *Carpe astem*. Convenience has nothing to do with it, but rather the patient gathering of food and knowledge. Supermarkets dumb us down, while foraging tunes us up, demanding that we be fully alert and engaged. 'Plants teach me,' says Robin. 'We've lost the natural granularity and *terroir* of place – of how this hawthorne community here can taste different to one literally twenty yards away.' Tastes are constantly changing too, since the ecosystem is always shifting. 'You can't dismiss a plant,' Robin explains; 'you've got to try it through the seasons.'

The sun has climbed high in the sky by now and the shingle is shimmering in the heat. My rumbling stomach reminds me that, delicious though they may be, a few sea vegetables don't quite amount to lunch. I muse on this as we walk back to the car park. Over the past few hours, Robin and I have been immersed in another world, far from the deadlines and pressures of modernity. As my mind wanders towards the interesting topic of what the local café may have on the menu, I feel I am fast-forwarding through time, from a distant existence when getting lunch was pretty much all one did. I'm grateful that I'll soon be able to satisfy my hunger with almost no effort on my part (visions of a Devon cream tea float across my mind), but I'm also aware of what I'll be missing by eating so effortlessly. The truth is that progress, however desirable, rarely comes without loss.

As we reach the car, I look back at the landscape I first glimpsed early this morning. The scene is as I remember it, yet somehow it feels different. I stare harder. The sea, beach, cliffs and trees all look the same, yet in just a few hours, I realise, my view has been transformed.

Blue Planet

> By diminishing nature we diminish ourselves.
> Wendell Berry[1]

Our relationship with nature is of course far from as harmonious as the one I've just described. We have long exchanged full immersion in nature for some degree of technical mastery over it, a bargain that has reduced our bond with the wild as well as its own natural richness. So extreme are the effects of our actions, indeed, that we have entered the Anthropocene, an age in which human activities are a decisive influence on planetary ecosystems.[2] Quite when the era started is a matter of debate: some suggest the first nuclear explosion, or Trinity Test, in 1945, while others favour the Industrial Revolution, when we first started pumping serious amounts of carbon into the atmosphere. Yet the global impacts of our activities go back much further than that.

Our journey out of Africa, which started 70,000 years ago, has spelled disaster for many of the creatures we've met along the way. Before our arrival in Australia 45,000 years ago, for example, the continent was home to a spectacular menagerie of twenty-four large animal species including giant koalas, marsupial lions and a two-and-a-half-ton wombat, all but one of which were wiped out within a few thousand years of cohabiting with humans.[3] North American bison were not the first beasts to be massacred on that continent either: of the forty-seven species living there when our ancestors arrived 14,000 years ago, thirty-four had vanished within 2,000 years. In total, of the 200 large mammals that roamed the earth at the start of our diaspora, one half were extinct by the time we started farming. As Yuval Noah Harari has remarked, our progress makes the human species 'look like an ecological serial killer'.[4]

Today it is farming, rather than hunting, that is our most lethal activity. By selectively domesticating certain plants and animals at the expense of others, we have drastically reduced the range and variety of wild species on earth and are now destroying the habitats of those not included in our domestic circle. Ecologists warn that we are on the brink of a sixth mass extinction that could prove as deadly as the one that did for the dinosaurs. A 2017 study by a team including US biologist Paul R. Ehrlich found that one third of vertebrate species were in decline.[5] All of those studied since 1900 had lost 30 per cent or more of their range, while 40 per cent had experienced severe population declines of 80 per cent or greater. As the authors observed, such figures are sure signs of future extinctions. Indeed, with the loss of some 200 vertebrates over the past century alone, the current rate of extinction over such a period is one hundred times greater than the average for the past two million years. The authors' conclusion was stark. We are, they said, in the midst of a 'biological annihilation' that represents 'a massive anthropogenic erosion of biodiversity and of the ecosystem services essential to civilization'.[6]

Biodiversity loss doesn't scare us as much as it should. We get upset about the plight of tigers or polar bears, yet few of us see their fate as directly linked to our own – we have, after all, got plenty of form in outliving majestic beasts. Yet the loss of species represents a threat to us potentially far greater than climate change. One need only reflect on Darwin's insight that everything in nature is connected. It's not just magnificent 'poster animals' that we need to worry about, but invertebrates, of which by far the largest group is insects. Most of us would probably cheerfully sign up to a world without wasps or mosquitoes, yet their absence could spell catastrophe, as Rachel Carson argued in her 1962 book *Silent Spring*. The *Spring* in Carson's title was *Silent* because the indiscriminate spraying of crops with DDT in the US after the Second World War wiped out the majority of farmland insects and thus most of the birds that lived on them.

Half a century after Carson's warning, 'Insectageddon' is going global. In 2017, a 27-year study carried out on nature reserves in Germany found that numbers of flying insects had plummeted by 76 per cent.[7] Habitat loss, pesticide use and climate change were cited as the most likely causes. The study confirmed what many in Europe had been casually observing for years: that windscreens once spattered with bugs while driving on summer nights were now spookily pristine. The implications of such a bugless world are sinister indeed. As well as being a vital part of the food chain, insects are key pollinators for many fruits and other food crops as well as virtually all wild plants. They are essential to natural life-cycles, breaking down animal and vegetable matter and recycling nutrients in the soil. Without them, the entomologist E. O. Wilson reckons, humans would last a matter of months.

With their favourite food disappearing, it's no surprise that birds are following suit. In 2017, the French Museum of Natural History reported that farmland birds in France had declined by 33 per cent since 1989, a finding mirrored by a UK Defra study reporting a 55 per cent loss of birds since 1970.[8] The global picture is similarly bleak: in 2018, BirdLife International reported that 40 per cent of the world's bird species were in decline, with 13 per cent under direct threat of extinction.[9] Of the most endangered, 74 per cent were directly impacted by agricultural intensification, identified as the main culprit in the worldwide decline of bird populations.[10] Less visible, yet no less ominous, is the plight of sea life: in 2018, the FAO reckoned that industrial trawling has led to nearly 90 per cent of the world's marine fish stocks being fully exploited, over-exploited or depleted, while warming oceans are leading to the widespread loss of coral.[11]

The millions of familiar creatures with which we coexist or unknown critters that may vanish before we ever clock their existence aren't just handy workers or essential links in the food chain; they are our greatest repository of intelligence about how to live on earth, collated over almost four billion years. All our foods and medicines come from nature and nobody knows what may yet be out there: chemicals extracted from sea slugs, for example, are currently being tested as a potential cure for cancer.[12] Biodiversity matters because it represents an interconnectivity that we don't fully understand. Like the Apollo 13 astronauts who resorted to using bits of spare kit to mend their stricken spaceship, we need to cherish our fellow cohabitants of earth not just because they represent the wonder of life, but also our best – indeed only – chance of sharing in its future.

It's not all about *us*, of course; on the contrary, plenty of people have pointed out that the planet would do very well without us. The paradox is that, without us, there would be no 'better' or 'worse': only humans invest the world with such meanings. Furthermore, despite our destructiveness, most of us have an innate love of nature. This was demonstrated in 2017 when David Attenborough's BBC series *Blue Planet II*, which featured heart-warming footage of surfing dolphins and snoozing whales, was followed by shocking scenes of a sea turtle snagged in plastic. The series had an immediate and dramatic impact, triggering urgent debate and prompting the UK government to set new targets for plastic use and major companies pledging to reduce theirs.[13]

Small is Beautiful

As *Blue Planet II* demonstrated, our relationship with nature is powerfully governed by emotion. We care more about birds than insects because our feathered friends are beautiful, possess the gift of flight and demonstrate behaviours, such as migrating thousands of miles, selflessly feeding their young and sometimes mating for life, that we find noble. Although a select group of humans are devoted to bugs, a far greater number of us are bird-lovers, yet, without insects, there would be no birds and no us.

As the tree-huggers among us testify, however, nature doesn't have to be cute and cuddly to inspire our love. Indeed, it needn't even be 'alive' in order to move us: many of us, for example, feel drawn to mountains or renewed just by being by the sea. In this respect, searching for images of nature on Google is revealing: of the first twenty that came up when I looked, fifteen featured trees and ten water, six were of mountains and five close-ups of plants. Of all twenty images, just one included a human (a silhouetted figure far off in the distance) and none featured animals at all.

Which is a bit strange, when you come to think of it. When we imagine nature, we tend to conjure up majestic scenes such as an ancient forest, some snowy mountains or a storm at sea. What we almost never picture is ourselves, or the microscopic beings that squiggle beneath our rose bushes or line the walls of our gut. Yet microbes (bacteria, viruses, protozoa, algae and fungi) are what connect us, directly and literally, to the natural world. More essential to life than insects, they animate all the mountains, forests, fish, roses and humans on the planet. Without microbes there would be no life on earth, which is why, if we really want to connect with nature, we need to start thinking small.

For those unversed in the power of exponentials, it can be hard to grasp quite how many microbes there are on earth. There are an estimated 10^{30} of them (that's a figure 1 with thirty noughts after it), one million times more than there are stars in the known universe.[14] Too small to be seen by the naked eye, microbes nevertheless constitute one half of all the living mass on earth. If you find that hard to digest, try swallowing the fact that those in your gut weigh around four pounds.[15] Microbes are absolutely everywhere: not just in laboratory Petri dishes or the mouldy yoghurt at the back of your fridge, but in rocks, soil, sea and air, on trees, flowers, birds and bees, on your dog's nose and your lover's lips, on the book, tablet or phone you're holding right now, in your mouth, eyes, skin and guts. Our bodies contain some 100 trillion of them, the cells of which outnumber our own by at least three to one.[16] With statistics like that, it's fair to wonder just how human we really are.

If microbes were half (or one trillionth) as lethal as we tend to think they are, we'd all be dead by now. The fact that we thrive alongside them – and they with us – suggests a very different picture to the one painted by our germ-phobic, hygiene-obsessed culture. Although some microbes, classed as pathogens, can indeed kill us with ruthless efficiency, the vast majority are not merely friendly, but essential to our existence.

Microbes are ubiquitous, so what are they actually doing? The answer is that they're doing just the same as you and me: trying to thrive in a complex, competitive world. They've been at it far longer than us too: the first earthly life forms, microbes are thought to have emerged around 3.85 billion years ago, when charged sea particles swallowed some mineral 'soup' belched forth by a hydrothermal vent (a volcanic fissure in the ocean floor) to form single-celled bodies known as archaea.[17] By eating the world's first meal, therefore, archaea, our common ancestors, kick-started life on earth, using chemical energy to process carbon, hydrogen, oxygen and nitrogen to form amino acids, the building blocks of life.

For a billion years or so, archaea ruled on our acidic, sulphurous planet, but around 2.7 billion years ago, some deadly rivals came on the scene. Cyanobacteria, or blue-green algae, began proliferating in the oceans, using solar energy to absorb hydrogen from water, expelling oxygen as waste. This process – a primal form of photosynthesis – transformed our planet. A highly promiscuous element, oxygen combined with everything in sight, not least with iron to form the world's first rust belts. For the archaea, this spelled catastrophe: since oxygen is a deadly poison to them, our most distant ancestors either died or disappeared underground. Their exile has proved permanent: when the oceans became saturated with oxygen, the gas began leaking into the air to create the Great Oxygenation Event, the basis of our modern atmosphere. Once oxygen levels had stabilised at 21 per cent around 0.9 billion years ago, the stage was set for complex life to evolve, and for animals like us to walk the earth.[18]

Most of us probably know from dimly recalled biology lessons that plants use sunlight and water to perform photosynthesis and that a key by-product of the process is oxygen. What fewer of us may realise is that the process depends on microbes. Chloroplasts, the sub-cellular organisms in plants that perform photosynthesis, consist of none other than our old friend cyanobacteria, still merrily chugging out O_2 into the atmosphere. Photosynthesis (from Greek *phos*, light + *sunthesis*, put together) remains the foundation of all complex life on earth, providing us with the food we eat and the air we breathe. In its modern form, it involves the conversion of sunlight into chemical energy, which is then used to combine elements from water and carbon dioxide to form carbohydrates, the basis of the food chain. Since this trick remains the sole preserve of cyanobacteria, algae and plants, it is to them that the rest of us earthlings (with the exception of the banished archaea) owe our lives. When Isaiah said 'all flesh is grass', he wasn't far wrong: we all depend on beings capable, sometimes literally, of making hay while the sun shines.

Apart from letting us breathe, cyanobacteria have one more trick up their tiny sleeves: the ability to fix atmospheric nitrogen. Nitrogen, as we saw in the first chapter, is a key nutrient for both plants and animals, yet exists mostly in air and must be fixed before it can be taken up by plants. Long before Messrs Haber and Bosch came along, cyanobacteria were performing this useful function, for example keeping flooded Asian rice paddies fertile from one century to the next.

We owe our lives to cyanobacteria, so why haven't more of us heard of them? One answer lies in the fact that they're so small that our awareness of them is only now growing, as advanced microscopy allows us to peer into their world. Just as space telescopes like the Hubble have expanded our knowledge of the universe, electron microscopes are transforming our understanding of the microbial realm. While some scientists gaze up at the stars and dream of colonising Mars, others point down at the earth beneath our feet and wonder what treasures lie buried there. At both ends of the spectrum, our comprehension of nature is undergoing a revolution.

Man and Cosmos

> Nature is a principle of change.
>
> Aristotle[19]

The fact that we in the West understand nature mostly by peering at it through instruments says a lot about our world view. The fact that it feels natural to us to observe nature from afar is, when you think about it, something of a paradox. If an act can feel 'natural' to us, that implies we're part of nature; yet if we can observe it from the outside, then surely we can't be? This conundrum has dogged Western thought at least since the days of *Gilgamesh* and lies at the root of our deepest dilemmas and greatest triumphs, culminating in the Enlightenment.

The seeds were sown, as ever, by the ancient Greeks. Aristotle was the first to question the principles of nature in what we would now recognise as a scientific manner, setting out to observe and explain natural phenomena. For Aristotle, Nature with a capital N essentially meant what we mean by it today: the sum of all plants, animals and objects in the natural world. The nature (with a small n) of such entities was determined by an inner blueprint that told them which form to assume. Nature was thus in a constant state of becoming, as living things obeyed their inner instructions to assume their proper form: acorns became oak trees, calves matured into cows, and so on. Since form followed function, fish had fins in order to swim, birds grew wings with which to fly, and so forth. Each species was thus ideally adapted to suit its habitat, creating a natural hierarchy that also conveniently provided everyone with lunch:'We must assume,' said Aristotle,'that plants exist for their [the herbivores'] sake, and that the other animals exist for the sake of man'.[20]

If Aristotle's ideas seem faintly familiar, it's because he effectively invented natural science, laying the foundations for the rationality of the Enlightenment. His idea that species perfected themselves according to their function and habitat is not, after all, far removed from Darwin. Aristotle objectified nature, yet for him humanity remained very much part of its order. For nature to become truly objectified, humans would have to be set apart, a process that started, as we have seen, with the Biblical myth of creation.[21] The Garden of Eden remains our seminal image of nature: a vision of earth in its pristine state, glorious enough to represent paradise itself.Yet man was never fully part of the image. Given dominion over his fellow creatures from the start, Adam enjoyed a privileged position that Christianity would elevate to full-blown transcendence, granting humans a soul and thus the prospect of joining their maker in heaven.

The repercussions of this cosmic shift still reverberate. Whether or not one believes the creation myth to be literal truth, the idea that humanity is somehow set apart from nature remains firmly fixed. It was a concept that dogged medieval and Renaissance scientists and theologians, as they tried to reconcile the *ménage à trois* between Man, Nature and God that placed man apart from nature, yet at the centre of creation. The crisis came to a head in 1543, when a Polish astronomer-monk by the name of Nicolaus Copernicus suggested that the planets revolved not around the earth, but around the sun. Copernicus' blasphemy cost him his life, yet it opened the fatal fissure in the Christian cosmos that signalled the start of the Enlightenment. This radical rearrangement of the scientific and philosophical furniture reached its early apotheosis with the publication in 1687 of Isaac Newton's *Philosophiæ Naturalis Principia Mathematica*, a monumental effort of synthesis in which the greatest scientist of his day combined the laws of gravity, motion and light to create a working model of the entire universe.

The *Principia*, as its title indicates, was a work of natural philosophy: a tradition of thought in which science, philosophy and theology were combined in the same ontological quest.[22] Its central message – that the universe could be explained entirely through mathematics – thus shook not just the scientific establishment but all of intellectual society. If the universe were really arranged according to predictable laws, that meant that man, although demoted from the centre of creation, could nevertheless hope to command it through logic and understanding alone. Never before had the world seemed so accessible to human reason, prompting the question of who this newly commanding creature – man – was himself.

For the French mathematician and polymath René Descartes, the answer was 'a thinking being'. Inspired by Newton's precise description of the universe to come up with his own for human experience, Descartes set out in his 1637 *Discourse on Method* to ask what, if anything, we could know for certain. We can't trust our senses, he argued, since they can be easily fooled: a stick held under water, for example, seems to be bent in the middle. We can't trust our reason either, since we know we're prone to error; we can't even trust our thoughts, since our dreams often appear real to us. What then, asked Descartes, *can* we trust? The answer was that, in the very act of asking such a question, we know ourselves to be asking it. 'In order to think,' Descartes said, 'it is necessary to exist.'[23]

Descartes' famous dictum '*Cogito, ergo sum*' (I think, therefore I am) gave birth to Western rationality. In Descartes' cosmos (henceforth known as Cartesian), man alone was conscious, while everything around him, including his own body, belonged to a mechanistic, material realm. Animals, without a soul or the capacity to reason, were reduced to mere machines. When observing a dog or monkey, said Descartes, there was nothing to indicate that their actions were not just the result of mechanical impulses, not unlike the clockwork figures popular at the time. 'Those who are cognisant of how many different automata or moving machines the ingenuity of men can make,' he wrote, 'will regard this [the animal's] body as a machine which, having been made by the hands of God, is incomparably better ordered.'[24]

By dividing nature into mind and matter, Descartes effectively placed man where God had once stood. His Cartesian universe was not one of wonder, but rather a rational gridded space whose three dimensions were described as intersecting x, y and z axes, all meeting at their respective point of zero. This remarkable abstraction, whose most visible legacy has been to give us the ubiquitous graph, transformed our concept of space and thus our understanding of reality.[25]

Mind over Matter

It is testimony to the power of Cartesian geometry and Newtonian physics that, three centuries after they were conceived, we were able to use them to put men on the moon. The Enlightenment gave us modernity, yet it also bequeathed us a splintered world in which we struggle to find ourselves. Natural philosophy has shattered into a thousand fragments, among which mathematics reigns supreme. Graphs and statistics dominate our thinking; only ideas expressed in numbers are truly valued. The Enlightenment gave us many gifts – science, knowledge and understanding – yet it also gave us thinking in silos.

Without disciplinary thinking, we should never have split the atom, discovered antibiotics or unlocked the secrets of DNA. Yet our technical prowess has outpaced our philosophical wisdom. Like Prometheus with his stolen flame, we've acquired godlike power without the chops to handle it. Nowhere is this imbalance more apparent than in our relationship with nature as made manifest by the way we eat. Descartes' machine-animals have leaped straight off the pages of his *Discourse* to form the centrepiece of our post-industrial plate.

Factory farming remains the most morally dubious outcome of our aloofness from nature. As Keith Thomas remarked in *Man and the Natural World*, it's hard to recapture the 'breathtakingly anthropocentric spirit' with which early modern thinkers seized upon Descartes' notion that animals were unable to feel pain.[26] Even John Locke admitted that it was easier 'to conclude all beasts were machines rather than allow their souls immortality'.[27] Today we know that, far from being senseless machines, farm animals are both highly intelligent and have sophisticated social and emotional lives not dissimilar to our own.[28] So what's our excuse now?

We don't have one, of course. Instead we have a massive case of cognitive dissonance, with a barely suppressed guilt complex to match. Although attitudes towards meat-eating are starting to shift in the West (in 2018 there were an estimated 600,000 vegans in Britain, up from just 150,000 in 2006), our mainstream food culture remains stubbornly carnivorous.[29] When UK branches of Kentucky Fried Chicken ran out of, of all things, chicken in 2018, punters were so enraged that many phoned the police, forcing several constabularies to inform the public that the temporary unavailability of their favourite snack was 'not a police matter'.[30]

Nature Abhors a Vacuum

While running out of chicken is inconvenient, compared to the serious threats posed by our poultry habit, it is mere chicken feed. In 2018, research led by Cardiff University Professor of Microbiology Timothy Walsh found that colistin, an 'antibiotic of last resort', was being routinely used in factory chicken farms in Russia, India, Vietnam and South Korea to boost the birds' growth. Users included major suppliers to (you've guessed it) KFC, as well as Pizza Hut and McDonald's. The team found that use of the drug was spreading rapidly, despite the discovery in 2015 of the colistin-resistant gene mcr-1 in some Chinese pigs. The finding caused widespread panic in medical circles, since the gene can be easily transferred: indeed, it has already spread in various forms to thirty other countries. When asked to comment on the use of colistin in chicken feed, Walsh didn't mince his words: he called it 'complete and utter madness'.[31]

Why are we risking our own lives to feed ourselves? The answer is that we expect food to be cheap. Thanks to industrialisation, we've forgotten what food really is: living emissaries from the natural world. By treating nature as a resource to be exploited, we have devalued it. Our dilemma is that, in order to live, we *must* manipulate nature, yet must seek to do so without diminishing it. Farmers have always manipulated nature, of course, but only recently has this interference threatened global ecosystems. This is, of course, partly to do with scale, but it is also to do with the nature of the disruption. As Essex University Professor of Science and Society Jules Pretty has argued, biotechnology is not necessarily bad *per se*: the selection of plant genes for more salt or drought resistance, for example, is merely a continuation of breeding practices that have been carried out for millennia.[32] Other 'first generation' forms of GM, however, which consist mainly of the breeding of crops to resist specific herbicides or pests, are a very different story.

One case in point is glyphosate, the toxic key ingredient in Monsanto's bestselling weedkiller Roundup. Initially hailed by farmers and scientists alike as a once-in-a-century discovery, the weedkiller was designed to be used on 'Roundup Ready' GM crops engineered to resist its deadly poison. After two decades of happily spraying their fields with the stuff, however, US farmers found to their dismay that the weeds were fighting back. The first appeared in a soybean field in Delaware in 2000, since when more than ten different sorts (with combative names such as horseweed, pigweed and giant ragweed) have emerged, affecting more than ten million acres of soybean, cotton and corn fields across twenty-two states. As their monikers suggest, these Roundup Ready weeds are no blushing violets: some grow up to seven feet tall and have stalks thick enough to damage farm machinery. Where did these monsters come from? The answer is that nature abhors a vacuum, and to horseweed, pigweed and friends vast tracts of monocultural farmland bereft of their usual competitors was an unmissable opportunity. As one weed scientist put it, it was a case of 'Darwinian evolution in fast-forward'.[33]

When working with nature, silver bullets are rarely the answer. Natural systems are inherently complex, maintaining balance through the principle of reciprocity: 'good' microbes naturally fight pathogens, natural predators eat pests, and plants exude phytochemicals to defend themselves.[34] Nature builds resilience through complexity, yet simplifying it has long been the aim of farming. Of the estimated 300,000 edible plant species on earth, just seventeen now provide 90 per cent of all our food.[35] Without agriculture, we wouldn't have sandwiches, cities or seedless grapes, but as we enter the urban age, all our past weeding and breeding is coming back to haunt us. Our food system is streamlined, efficient – and vulnerable. Our grandmothers would have called it putting all our eggs in one basket.

Complete and utter madness would be another fair description. You don't have to be a fortune teller, microbiology professor or even a granny to see that our current approach to food is fraught with risks. Yet we persist. One reason why is inertia: we have, after all, been farming for millennia and old habits die hard. Another is power: the food system is increasingly controlled by global conglomerates with a vested interest in maintaining the status quo. But arguably the key reason is that our Cartesian world view makes us believe that we can master nature through technology. Just as economic growth (represented by a pleasingly upward-sloping line on a graph) is now synonymous with progress, ever-increasing crop yields have become our only benchmark for judging whether or not farming is working.

We're engaged in an arms race with nature in which logic suggests there can only be one winner. So what are our options for feeding ourselves in the future? Or, to put it another way: what sort of future relationship with nature do we want?

Two Schools

Whether mowing down bison from moving trains, destroying rainforests or fishing bluefin tuna to the brink of oblivion, we have consistently failed to live up to the role of caring custodian to which our own myths have appointed us. Today, there is growing consensus that we need to do things differently, yet little agreement as to the best way forward.

On the question of how we should feed ourselves, there are broadly two schools of thought. The first, and by far the most powerful, is the industrial lobby. This group argues that we need to accelerate our mastery of nature, getting ever smarter in our use of fertilisers and pesticides, genetically modifying plants and animals and increasing efficiencies of production wherever possible. We need, in other words, to stick to Plan A, only do it a lot better. The more we can concentrate our destructive human activities, the thinking goes, the more nature we can preserve by releasing it back to wilderness. The second group is the organic lobby, whose message is, as you might expect, pretty much the opposite of the above. Pointing to the catastrophic externalities of industrial agriculture, this group argues that, instead of trying to master nature, we need a Plan B, based on developing and rediscovering ways of working with the natural world to build greater soil fertility, diversity, complexity and resilience into the ecosystems that feed us.

What is striking about these two schools of thought – apart from their unhelpful polarity – is how differently they view the entity that both set out to save: nature. For the industrialists, nature is split into two halves: a domestic part that we should exploit to its limits, and a wild one that we should leave well alone. Access to wilderness, for this group, is a luxury we can no longer afford. For the organic school, on the other hand, nature is a continuum whose wild and domestic parts intermingle. Closeness to wildness is part of this group's aim, not least by incorporating more of it into farming.

Lurking beneath this philosophical divide, you may perceive, lies another one, over the question of whether humans are capable of looking after nature. While the industrialists argue that we can't be trusted and must therefore withdraw from wilderness, the organicists contend that greater engagement with the natural world fosters better behaviour. The important thing to consider is which approach holds the most promise of a good life in the future – for both humans and non-humans.

Nature and Culture

The French anthropologist Philippe Descola argues that first we need to recognise that our very concept of nature is a cultural construct.[36] This is hard for us to accept in the West, says Descola, since we believe that we view nature with objective, scientific clarity, which in turn makes us see nature and culture as mutually exclusive opposites. Nature, for us, is a sort of neutral backdrop against which various cultures play out. This view led early-modern European voyagers to dismiss indigenous peoples who invested nature with spirits as mere savages mired in so much mumbo-jumbo. The idea that their *own* view of nature, with its abstract space and machine-animals, might itself be cultural never occurred to them.

Armed with their navigational instruments, seventeenth-century Western explorers believed that only they saw nature as it truly was, says Descola, yet one could argue that it was the other way round. The Western division of nature into wild and domestic, for example, has no direct equivalent elsewhere. In traditional Japanese culture, for instance, a key distinction is made between mountains and plains, the former being uninhabited and vertical, while the latter are crowded and flat. Yet both spaces are held as equally sacred: the mountain gods are said to descend every spring to spend the summer in the rice paddies, before returning home in the autumn. In this way, says Descola, 'The distinction between *yama*, the mountain, and *samo*, the inhabited space, signals not so much a reciprocal exclusion but, rather, a seasonal alternation and a spiritual complementarity.'[37] A similar relationship can be found in Chinese Daoism, where mountains are seen as the dwelling places of the immortals, providing a spiritual counterpart to the practical life of the plains.

Such complementarity is not unknown in the West: in ancient Greece, for example, wild woods were seen as sacred places where men went to spiritually renew themselves and to develop their strength and agility through hunting. Although distinct from cultivated land, the forest was valued as a source of food, and, since all meat was sacrificed and served alongside farmed produce such as toasted barley and wine, wild and domestic nature were joined in the ritual of the meal. To the Greeks, wilderness was thus seen as a necessary adjunct and counterbalance to civilisation.

This positive view did not, as we have seen, survive into Roman times. For the tidy-minded Romans, wilderness represented little more than territory yet to be tamed and brought within the civilised realm. The forests of Germania were thus a provocation, not just because their savage inhabitants were so annoyingly good at fighting, but because they represented a threat to civilisation itself.

Despite their dread of the Germans, the Romans felt a grudging respect for their strength and virility, holding the purity of their lives in the wild to be largely responsible. This sort of urban reverse snobbery, replete with its latent self-loathing, would resurface centuries later in the full-blown guise of Romanticism. As cities like Paris and London swelled during the eighteenth century and agronomists such as 'Turnip'

Townshend trumpeted the benefits of land enclosure, forest clearance and cattle fodder in order to feed them, some began to question whether civilisation itself was worth the candle. The principal naysayer was Jean-Jacques Rousseau, whose 1750 *Discourse on the Sciences and Arts* was a scathing attack on the gilded Parisian society of which he himself was a prominent member. It was all very well pontificating in salons, said Rousseau, but in truth all that civilised society really did was to cloud one's judgement about what really mattered in life. If mankind wanted to recover its dignity, its only hope was to return to its natural roots:

> We cannot reflect on the morality of mankind without contemplating with pleasure the picture of the simplicity which prevailed in the earliest times. This image may be justly compared to a beautiful coast, adorned only by the hands of nature; towards which our eyes are constantly turned, and which we see receding with regret.[38]

By suggesting that nature, not culture, was the source of human virtue (thereby placing the 'noble savage' above civilised man), Rousseau effectively flipped Enlightenment values on their head. Science and knowledge had their uses, he said, but only if they were grounded in earthbound morality. 'The sciences, letters and arts,' he wrote, '. . . wrap garlands of flowers around the chains that weigh people down.'[39] Far better to live in a rustic hut than submit to the 'loathsome and deceptive conformity' that prevailed among the 'herd of society'. Compared to such artificiality, 'primitive people' who had never known civilised life were blessed. 'I dare not speak about those happy nations,' said Rousseau, 'who did not even know the name of the vices we struggle to suppress.'[40]

Call of the Wild

> In wildness is the preservation of the world.
> Henry David Thoreau[41]

Ridiculed by his contemporaries, Rousseau nevertheless emerged as the standard-bearer of a band of poets and adventurers for whom the call of the wild was becoming irresistible. As Edmund Burke noted in his 1757 *Philosophical Enquiry into the Origin of Our Ideas of the Sublime and Beautiful*, people felt drawn to soaring peaks, yawning chasms and gushing cascades, because their dramatic scale and natural majesty inspired awe and thus allowed them to dream of the infinite. Here was a very different kind of infinity to the gridded certitudes of Cartesianism. The call of the sublime was, in many ways, a riposte to such calculated precision: a direct appeal to the human heart and soul. Like Caspar David Friedrich's *Wanderer above the Sea of Fog*, Romantics didn't dream of mastering wilderness, but rather revered its divine mystery.

The sooty imprints of industrialisation turned the yearning for wilderness into a clamour, nowhere more so than in the US, where there was still (at least in theory) plenty of it to go round. The mood was captured by Ralph Waldo Emerson in his 1836 essay *Nature*, in which he spoke of the deep need for spiritual renewal through the embrace of 'essences unchanged by man; space, the air, the river, the leaf'. Emerson thought that mankind's lost connection with nature was akin to a sickness. 'Our age is retrospective,' he wrote. 'The foregoing generations beheld God and nature face to face; we, through their eyes. Why should not we also enjoy an original relation to the universe? . . . The sun shines to-day also.'[42]

For Emerson, nature was everything good: divine and beautiful, a teacher and healer, the source of all nourishment and the balm that soothed every woe. 'The lover of nature,' he wrote, 'is he whose inward and outward senses are still truly adjusted to each other . . . His intercourse with heaven and earth becomes part of his daily food.'[43] For those in tune with nature, transcendence could be achieved simply by looking up at the sky. 'If a man would be alone,' wrote Emerson, 'let him look at the stars.'[44] Yet it was among trees (in the 'plantations of God') that people found their greatest succour. When we are in the forest, said Emerson, we experience the 'occult relation between man and the vegetable', so regaining the wonder of childhood that is the basis of all 'reason and faith'.[45] Nature, in summary, was the antidote to civilisation. 'To the body and mind which have been cramped by noxious work or company,' Emerson declared, 'nature is medicinal.'[46]

Nature established the great outdoors as Americans' spiritual home. To Jefferson's citizen-farmer, Emerson added the rugged frontiersman, whose bond with wilderness lent him a natural nobility. The essay was eagerly received among leading intellectuals of the day, but it would fall to a young protégé of Emerson's by the name of Henry David Thoreau to truly popularise his vision. In 1845, the twenty-seven-year-old Thoreau, who had been acting as Emerson's assistant at his house in Concord, Massachusetts, left his master's home to spend two years in a rustic hut on the shores of Walden Pond. Thoreau's memoirs, published in 1854 as *Walden, or Life in the Woods*, were destined to outshine even his mentor's famous essay. Part diary, part almanac and DIY manual for the would-be hermit, *Walden* detailed Thoreau's everyday existence – tending his bean patch, listening to birdsong, or taking his daily plunge in the eponymous pond ('a religious experience') – alongside various philosophical musings. Despite making frequent trips into nearby Concord for supplies or just for company, Thoreau echoed Rousseau's critique of the civilisation he had theoretically left behind: 'Most of the luxuries and many of the so-called comforts of life are not only not indispensable,' he wrote, 'but positive hindrances to the elevation of mankind.'[47]

Despite receiving a drubbing from critics, many of whom saw Thoreau's 'experiment' as mere quixotic self-indulgence, *Walden* won the popular vote and was soon the unofficial handbook of a nascent environmental movement. Enter John Muir, a rough-hewn Scot and fervent Emerson fan whose 1867 apprenticeship into wilderness was rather more thorough than Thoreau's, consisting, as it did, of a thousand-mile trek from Indiana to Florida, along what Muir described as the 'wildest, leafiest, and least trodden way I could find'.[48] Muir's real epiphany, however, came the following year, when he set eyes on the landscape that would define his life, Yosemite Valley in the Sierra Nevada. 'No temple made with hands can compare with Yosemite,' Muir gushed, '. . . as if into this one mountain mansion Nature had gathered her choicest treasures, to draw her lovers into close and confiding communion with her.'[49]

Muir moved into the valley, taking various casual jobs and building himself a log cabin over a stream. Increasingly blending into the landscape he so adored, Muir was soon a landmark in his own right, receiving a visit from no less an eminence than Emerson himself in 1871. Yet, despite feeling utterly at home in the wild, Muir gradually became convinced that Yosemite should be protected from all human influence, including his own. He began campaigning for the valley to be made into a national park, receiving an unexpected boost in 1903, when Theodore Roosevelt asked to go camping with him (in a *Boys' Own* trip echoed a century later by Bear Grylls' rather less rugged adventure with Barack Obama).[50] After sleeping out under the stars and waking up under a blanket of snow, the president declared himself a convert. In 1905, Yosemite was declared a national park, to be forever preserved in all its pristine glory.

The Trouble with Wilderness

The invariable mark of wisdom is to see the miraculous in the common.
Ralph Waldo Emerson[51]

The trouble was that Yosemite was anything but pristine. As Muir himself had acknowledged in his early writings, it was home to the Ahwahneechee people, whose deep bond with their landscape the Scot had noted, not without a hint of envy. One incident in which he ran out of bread on a camping trip brought home to Muir just how much he had to learn from the locals. 'The Indians put us to shame, as do the squirrels,' he wrote, 'starchy roots and seeds and bark in abundance, yet the failure of the meal sack disturbs our bodily balance, and threatens our best enjoyments.'[52]

While admiring the Indians' superior bushcraft, Muir increasingly baulked at the idea that his beloved valley was at least partly their creation. Yet the lush green meadows that carpeted the valley floor – so numinously portrayed in the canvasses of Albert Bierstadt – were the result, not of nature's artistry, but of the Indians' habit of periodically burning the forest in order to improve their hunting grounds. With the sanctification of Yosemite, however, such inconvenient facts had to be swept under the greensward: the Ahwahneechee were carefully airbrushed out of the picture.

With his unique blend of craggy woodsman and mystic poet, Muir fused environmentalism and Romanticism into a seamless whole. His carefully constructed fantasy of pristine wilderness completed what the

Old Testament had begun, placing humanity firmly outside nature's frame. This latest expulsion from Eden (a sort of self-imposed Second Fall) denied humans any place in earthly paradise. From now on, the only 'true' nature would be the sort untouched by human hand. This idea, as William Cronon argued in his essay 'The Trouble with Wilderness', took particular hold in America, where the frontier myth, with its vast unexplored horizons, seemed emblematic of a finer, vanished past against which modernity could only appear tawdry and tame. Wilderness, said Cronon, thus became 'the natural, unfallen antithesis of an unnatural civilisation that had lost its soul'.[53]

While English Romantics like Wordsworth and Coleridge had to make do with the miniaturised grandeur of the Lake District, Emerson, Muir and Co. had an entire continent on which to project their dreams. Vastness shaped American Romanticism, distracting US environmentalism from its key task of asking how man and nature might coexist. By worshipping wilderness at the expense of domestic nature, Cronon argued, American Romanticism induced a form of self-loathing that made people evade responsibility for the lives that they really led.

While few would disagree with the need to preserve wilderness, Cronon said, it is vital that we recognise that all landscapes form a continuum that is our home. Only when we can marvel at a humble shrub in a garden as we do a giant redwood in the forest will we have understood where our true place in nature lies. Thoreau was right: wildness (as distinct from wilderness) *is* the preservation of the world, yet we don't have to hike high in the sierra to find it, since it is all around us, in our cities, parks, homes, gardens and even in our own bodies.

Home Economics

> In human culture is the preservation of wildness.
> Wendell Berry[54]

For the US farmer-writer Wendell Berry, learning how to coexist with nature is 'the forever unfinished lifework of our species'.[55] The hardest part, he argues in his essay 'Preserving Wildness', is that we must decide our own limits:

> Humans, like all other creatures, must make a difference; otherwise, they cannot live. But unlike other creatures, humans must make a choice as to the kind and scale of the difference they make. If they choose to make too small a difference, they diminish their humanity. If they choose to make too great a difference, they diminish nature . . . Nature, then, is not only our source but also our limit and measure.[56]

Our key problem, Berry believes, is that we are unsure of where we stand in relation to nature. It can sometimes sound, he says, 'as if the natural and the human were two separate estates, radically different and radically divided'.[57] Yet a moment's thought should tell us that the two are inextricably linked. Pure nature, for example, is generally inhospitable to humans – we couldn't survive in the Arctic without some form of protection, or wander for long in the forest without facing a grizzly bear – yet to exclude nature from our lives completely would be just as fatal. Wherever we live, we need both nature *and* culture: the trick lies in blending the two. That, in essence, is what all traditional food cultures evolved to do.

There are no generic solutions to this, Berry argues, because nature is infinitely diverse, so our responses must be equally so. This is why trying to outsmart nature is so misguided. Take, for example, that 'miracle of modern science' the Holstein cow, bred to produce 50,000 pounds of milk a year. Such productivity seems wondrous until you factor in all the downsides: the fact that the cow may have trouble walking or be too delicate to go out in the rain, or that it eats bushels of grain grown with gallons of 'cheap' oil, using methods that deplete the soil or put farmers out of business.

We forget, says Berry, that the source of all flourishing is wildness: 'A forest or crop will be found to be healthy precisely to the extent that it is wild – able to collaborate with earth, air, light and water in the way common to plants before humans walked the earth.'[58] To live in balance with nature, we must therefore 'find some peace, even an alliance, between domestic and wild.' This means that we need cultures that teach us how to behave: 'It is more important than ever that we have cultures capable of making us into humans – creatures capable of prudence, justice, fortitude and temperance, and the other virtues. For our history reveals that, stripped of the restraints, disciplines, and ameliorations of culture, humans are not "natural", not "thinking animals" or "naked apes", but monsters.'[59]

Living in harmony with nature, in essence, is the wisdom of local tradition: 'The only thing we have to preserve nature with is culture, the only thing we have to preserve wildness with is domesticity.'[60]

Enclave Thinking

In the 200,000 years since *Homo sapiens* first walked the earth, we've tried umpteen ways of living with nature, some more successful than others. Success is a relative term, of course: whether you consider the Romans or the Mbuti to have made a better fist of it, for example, depends on your criteria. If they are stonemasonry, oratory and militarism, then the Romans are streets ahead; if empathy, equality and ecological longevity, the Mbuti clearly have it. Learning to balance nature and culture, as Berry noted, is our forever unfinished human task.

Today, on our shrinking, sweltering planet, our diverse approaches are starting to merge. As loggers, miners, ranchers and frackers drive out the last of the hunter-gatherers, our human story is coming full circle. How to live with nature is now a truly global question. Whatever the relative benefits of our options, one thing is clear: it will take far greater effort and imagination to move to Plan B than to stick with Plan A, for the simple reason that the latter represents the status quo. Plan B (going steady-state organic) would require us to change not just the way we eat, but how we live. It would mean swapping our consumerist fixes for less turbocharged pleasures, such as those gained from contact with nature (from which Plan A seeks to exclude us). Plan B seeks to replace the thrill of, say, upgrading our phone with the satisfaction of growing our own tomatoes. While few of us would willingly make such a swap, equally few of us are in a position to judge, since most of us now own a smartphone, but vanishingly few of us in the West grow our own vegetables, the reverse of the situation two generations ago. While nobody is suggesting that we all become farmers or, heaven forfend, give up our smartphones, Plan B will clearly involve shifting the sources of our joy.

Such change is already happening. There are plenty of those who, like rooftop farmer Ben Flanner, have realised that they prefer working with nature to sitting behind a desk. Indeed, as rural life is destroyed across the globe, increasing numbers of people are seeking to reinvent it. In both the UK and US, for example, independent cheese makers have seen a remarkable resurgence, with more than 700 farmhouse cheeses being made in Britain in 2016 and nearly 1,000 in the US.[61] The British rural population is also projected to increase by 6 per cent over the next decade, with many new farms and food businesses being set up.[62] In this sense, the food movement is like a deep ocean current: invisible on the surface, yet steadily gathering in pace and strength to the point where it can effect real change. Alongside parallel movements focused on sharing, repairing and craft, it suggests how we could indeed build new lives based on more lasting, creative pleasures.

Could such a vision scale up? For the industrial lobby, the answer is an emphatic no. Apart from arguing that we need to retreat from nature, this group argues that only Big Ag can feed the world. But is that really true? Certainly, if nothing else changes – if we continue to expect to pay next to nothing for our food – then no other food system could cope. If, on the other hand, we were to start valuing food again while moving to a zero-carbon economy, then a different food system would not only be possible, but necessary, since the industrial agri-food model would become untenable. Such an outcome is inevitable at some point in any case, since Big Ag will eventually run out of gas.

Two important, yet overlooked, insights emerge from this. The first is that those who argue for industrial agriculture fail to see that eating and living are inseparably linked, and that transformative change can flow in both directions. The second is that the claim that only Big Ag can feed the world ignores the possibility that we might live differently in the future, and is thus based on a false assumption. The organic alternatives are undoubtedly more complex and hard to envisage, yet that doesn't mean they couldn't work.

Feeding the World

In 2017, the UN Food and Agriculture Organization published a report called *The Future of Food and Agriculture* which predicted that by 2050 production of food, feed and fuels would need to increase by 50 per cent.[63] Citing climate change, urbanisation, land scarcity, ecological damage, conflict, waste, pests and diseases as just some of the challenges we face, the report stated that: 'High-input, resource-intensive farming systems, which have caused massive deforestation, water scarcities, soil depletion and high levels of greenhouse gas emissions, cannot deliver sustainable food and agricultural production.'[64] The report concluded that 'Business as usual . . . is not an option.'[65] Instead, what is needed is 'a transformative process towards holistic approaches, such as agro-ecology, agro-forestry, climate-smart agriculture and conservation agriculture, which also build upon indigenous and traditional knowledge'.[66]

By their very nature, such approaches are far more diverse than the one-size-fits-all NPK model. They can involve practices such as no-till (the avoidance of ploughing in order to reduce soil disturbance), green manures (the sowing of pulses and legumes to build up nitrogen in the soil) and herbal leys (temporary pastures alternately sown with other crops to restore soil fertility, sometimes combined with mob grazing, in which cattle are moved from one part of a field to another in order to mimic the movement of wild herds). They might also include various forms of permaculture: approaches to farming that seek to maximise natural synergies by mimicking nature. This can be done through the nurturing of biological complexity, the juxtaposition of habitats, mixed cultivation of trees, shrubs, pasture and crops and the capturing and conservation of nitrogen, energy and water, through methods such as the building of sand dams.[67] According to the report, such approaches have already been adopted on 117 million hectares of arable land world-wide, around 8 per cent of the world's total, with the highest adoption levels in Australia, Canada and South America.[68]

Unlike the ready-to-use methods of high-input farming, the FAO acknowledged, such practices are necessarily custom-made. Their success therefore depends on a raft of support initiatives – good education, reduced inequality, access to land, the empowerment of women, lower migration and rural investment – all of which need to be backed by global governance with the strength to transcend the usual divides between developed and developing nations. It is vital to have some form of shared vision to make all this happen, such as the United Nations' *2030 Agenda for Sustainable Development*, adopted in September 2015. 'On the path to sustainable development,' the report declared, 'all countries are interdependent.'[69]

The closest thing we've got to a shared global roadmap, the UN's *2030 Agenda* is an historically significant document, as full of hope as it is fragile. The trillion-dollar question is, of course, whether or not it can be made to work. Aside from the pivotal issue of whether or not the global governance it cries out for will ever materialise – let alone with sufficient teeth to force global corporate power into submission – is whether or not the agro-ecological methods it advocates could really feed the world. Since numbers tend to count in such debates, it's time to do some crunching.

One of the most painstaking recent attempts at such an exercise has been made by British dairy farmer and journalist Simon Fairlie. In his book *Meat: A Benign Extravagance*, Fairlie admits, somewhat wistfully, that his research didn't consist of 'splodging around farms in wellington boots', but instead involved the forensic examination of the plethora of data published on the subject of agriculture, which he found to be full of miscalculation, obfuscation and blatant 'mistruths'. Fairlie addresses the question so often overlooked in the how-to-feed-the-world debate: diet. He opens with a quote from the English poet Percy Bysshe Shelley, who in 1813 lamented the great 'waste of aliment' caused by the use of 'the most fertile districts' for the raising of livestock, just so that the wealthy could indulge in their 'unnatural craving for dead flesh'.[70] Shelley reminds us, said Fairlie, that ecological capacity and social equity are indelibly linked, and that nothing connects them more powerfully than meat.

The global rise in carnivorousness and the damaging downsides of industrial livestock production have led many to conclude that veganism is the only diet appropriate to our age. Actually, says Fairlie, it's a bit more complicated than that. While factory farming is 'unambiguously iniquitous' and the feeding of grain to animals 'one of the biggest eco-logical cock-ups in modern history', it doesn't follow that eating meat is in itself bad.[71] On the contrary, providing one has no moral objections to eating animals so long as they are raised humanely, there are powerful reasons for doing so since, when practised correctly, it allows us to pro-duce more and better food from the same resources. We need only to remind ourselves of why we humans domesticated farm animals in the first place: because they can eat what we can't. When fed on grass or food scraps as they always have been in traditional mixed farming sys-tems, they provide a nutritious source of food as well as a steady supply of valuable manure. Animals have always been integral to traditional farming, since they help to maximise local resources while returning nutrients to the land, thus creating highly efficient local ecosystems.

Such practices wouldn't have survived so long if they didn't make ecological sense; the question is whether they have any place in the modern world. To answer that, said Fairlie, we might look at how a nation like Britain could feed itself according to four different scenarios: with or without livestock and with or without chemicals. If we were to allocate everyone a ration equivalent to 2,700 calories a day based on cereals, potatoes, sugar, milk, meat, fruit, vegetables and beer (but with less meat and dairy than we currently eat and substituting peas and rape oil in the vegan case), then the headline is that Britain could easily feed itself.[72] Despite the caveat that such a system would require us to radi-cally alter our diet and behave like model citizens, the figure neverthe-less reveals a high latent capacity. Indeed, all four scenarios require less land than the 18.5 million hectares that we currently farm.

Of the four models, said Fairlie, organic-with-livestock struggles the most, requiring 8.1 million hectares of arable land and 7.8 of pasture to feed the nation, leaving just 2.6 million hectares of farmland spare. This kind of farming requires two hectares of land (one of arable and one of pasture) to feed 7.5 people, while the chemical-vegan option – the model that feeds us most easily – feeds as many as twenty from just one hectare of arable, leaving a whopping 15.5 million hectares of farm-land free; land that could thus be returned to wilderness. Of the two other models, organic-vegan fares best, feeding eight people per hectare to leave 11.2 million hectares of land unused, while the chemical-with-livestock scenario requires 2.5 hectares of land to feed fourteen people, leaving just 7.6 million hectares spare.[73]

Animal Farm

Apart from the obvious takeaway – that our diet and agricultural methods affect the amount of land needed to feed us – such exercises, as Fairlie admits, merely scratch the surface of the subject's true complexity. The organic-vegan approach, for example, may seem the most ecologically beneficial, yet in reality it would struggle to recycle nutrients, a function traditionally performed by animals. On organic mixed farms, the sowing of nitrogen-fixing leys such as clover is highly efficient, since it feeds cows while employing the animals to fertilise the fields on which they munch.

Scaled up, says Fairlie, the potential benefits of raising livestock by such traditional methods are huge. By feeding food waste to pigs, for example (an ancient practice banned in the EU after the 2001 foot and mouth crisis), we could produce 800,000 tonnes of pork a year in the UK, equivalent to one sixth of our total meat consumption.[74] And we shouldn't dismiss the idea of sowing some feed crops either, especially where they grow better than human food crops. Since animals are less fussy eaters than we are, they make far better use of what we grow: for example, we might waste half a field of potatoes through storage losses, rejects and peelings, while a herd of pigs would happily scoff the lot. Cattle will also gladly graze on harvest stubble, which represents a staggering amount of potential fodder: one 1999 study by the US food analyst J. G. Fadel estimated that global residues from just five major crops – wheat, rice, barley, maize and sugar cane – could provide cows with all the energy and one third of the protein they needed to provide the entire world supply of milk.[75]

Since animal-derived foods are on average 1.2 times as nutritious as vegetables and grain, Fairlie argues, farming animals as part of a plant-based approach makes perfect sense.[76] Dedicated carnivores shouldn't dash out and light up the barbie quite yet, however, since these findings are no excuse for gorging on steak. On the contrary, says Fairlie,

'Nowhere in this book do I make the case for eating lots of meat, because there isn't one. Meat is an extravagance.'[77] There *is*, however, he says, a convincing case for what he calls 'default livestock', which is to say, rearing animals as the 'integral co-product of an agricultural system dedicated to the provision of sustainable vegetable nourishment'.[78] Animals, in other words, could be reared much as they were in the past, mostly on nutrients – grass, food scraps and crop residues – that we would otherwise waste.

While such an approach would mean radically reducing the amount of meat and dairy we consume in the West, we needn't give up bacon and cheese altogether. Fairlie calculates that, practised globally, 'default livestock' would provide around half the meat and dairy we currently consume: around 18 kilos of meat a year per person (amounting to 350 grams a week) and 39 kilos of milk – equivalent to a weekly ration of 1.33 pints of milk or 75 grams of cheese.[79] Although this is not a huge amount, says Fairlie, it would effectively be free, since it would come from surpluses generated by the growing of grains and vegetables.[80]

There are, of course, other externalities to eating meat, not least the gasses emanating from both ends of Ermintrude and Co. Methane is an invisible menace common to all ruminants, estimated by the FAO to account for 5.4 per cent of all anthropogenic greenhouse gas emissions. Yet reducing such emissions isn't as straightforward as it may at first appear. While removing domestic bovines from the landscape may seem an obvious move, it would also come at an ecological cost. As regenerative ranchers Allan Savory and Tony Lovell have shown, the properly managed grazing of cattle on marginal grasslands can reverse desertification as well as act as an overall carbon sink.[81] The removal of all livestock from grassland also raises the question of what would occupy all the abandoned territory. If wild ruminants were to reclaim the world's 38 million square kilometres of pasture at the rate they've re-inhabited parts of the Serengeti, Fairlie reckons, they would produce an estimated 52 million tonnes of methane, five eighths the amount emitted by the current domestic herd.[82]

While this would clearly represent a major reduction, it raises the question of how, if we were to go vegan, we would handle the wild game that would naturally proliferate on our rewilded farmland. If we decided to hunt and eat the beasts in order to control their numbers, we would find ourselves back where our ancestors started millennia ago. If we chose not to eat the animals on ethical or other grounds, however, we'd be wasting a valuable food resource, for which we'd need to compensate by growing more grains and vegetables, foods that already contribute 17 per cent of our total methane emissions. We might want to avoid growing more rice, however, since its production, kilo for kilo, can emit up to four times more methane than that of milk.[83]

The fact that there is no ideal, methane-free solution to feeding ourselves shouldn't surprise us: we must use nature in order to live, after all, and some emissions are clearly part of the equation. What we *can* do, however, is seek to minimise our agricultural impact, while sharing our resources equitably, objectives that 'business as usual' will never deliver. Such aims suggest we should adopt a more integrated, low-input, vegetable-based existence – a sort of vegan–with–benefits lifestyle, if you will. To help the more committed carnivores among us to cut down on meat, we might start by internalising the true costs of factory farming. Such sitopian economics would naturally encourage a shift to plant-based organic production, but could we really feed ourselves without recourse to NPK?

Default Organic

The good news is that, although we couldn't feed ourselves entirely organically quite yet, we could get pretty close. This is according to a major study led by Dr Adrian Müller of ETH Zurich's Department of Environmental Systems Science, published in *Nature Communications* in 2017.[84] The study took a systems approach to the question of how we might feed the world in future, taking into account a number of variables such as diet, food waste and the likely effects of climate change. Using the FAO's data as their baseline (which call for a food supply of 3,028 calories per capita by 2050, assuming 30–40 per cent waste), the authors examined a range of agricultural approaches, together with their likely impact on greenhouse gas emissions, water and land use, deforestation, soil erosion and so on. Apart from confirming that organic farming is radically less harmful than conventional, the study found that, were we to halve our food waste and stop growing specialist animal feed (thus adopting Fairlie's 'default livestock' approach), we could feed ourselves 80 per cent organically using no more farmland than would be needed for the status quo.

As the authors point out, this figure could be further increased were we to invest in research to improve organic varieties and methods. Such improvements would almost certainly narrow the yield gap between chemical and organic crops – the difference in harvest yields, assumed by the report to be 8–25 per cent.[85] A 2014 study by a team from the University of California concurred, reckoning that, by using better organic varieties and methods such as intercropping (planting more than one crop in the same field), many such yield gaps would effectively disappear.[86]

It is no accident that such yield gaps tend to be highest in industrialised nations where vast amounts have been spent on developing high-yield crops that depend on chemical inputs. By contrast, a 2002 Essex University study of agricultural projects in the developing world led by Professor Jules Pretty found that such yield gaps in the Global South were radically reversed. Of the 208 projects studied across fifty-two countries – a sample involving 9 million farms and nearly 30 million hectares of land – Pretty and his team found that, where farmers had converted from conventional practices back to organic and ecological ones, yields had increased, on average, by a remarkable 93 per cent.[87] Such methods, the authors pointed out, tend to be more productive in such regions, since they are more suited to the smaller scale, lower cost, mixed production typical there.

The potential benefits of adopting such an approach worldwide are beyond measure. We could reduce water and energy use, greenhouse gas emissions, pesticides and pollution, and could reverse some of the species extinctions currently under way. If we're really serious about surviving the coming century with the world as we know it even half intact, then going 'default organic' is a no-brainer. Such an approach is probably our only chance of preserving wildness. As Michigan University paleo-ecologist Catherine Badgley has pointed out, much of the world's biodiversity exists in close proximity to farmland, so maintaining pockets of wilderness surrounded by oceans of pesticides won't work. 'If we simply try to maintain biodiversity in islands around the world,' she warned, 'we will lose most of it.'[88]

In 2019, the EAT-Lancet Commission's report *Food in the Anthropocene* confirmed such findings, concluding that the way we eat now poses an existential threat to both humans and planet.[89] Co-authored by thirty-seven global experts across food, health and farming, the report declared that nothing less than a 'Great Food Transformation' was needed, proposing the first science-based global diet aimed at addressing the dual threats of poor nutrition and ecological collapse. As well as food waste being halved, said the report, the amount of meat and sugar consumed globally would need to more than halve, with far greater reductions in wealthy nations, including an 84 per cent decrease in the amount of red meat eaten in the US and one of 77 per cent in Europe. To provide such a diet sustainably, a new agricultural revolution was also needed, involving the worldwide adoption of agro-ecological practices and 'sustainable intensification' to avoid the need for more agricultural land. To be effective, the report added, such measures would need to be backed up by strong and coordinated global governance to protect land, forests and oceans.

Despite objections that its proposed global diet failed to take sufficient account of cultural diversity, the EAT-Lancet report was welcomed by those who have long argued that we must place food back at the heart of our thinking. The idea was powerfully reinforced later that year by the IPCC's report *Climate Change and Land*, co-authored by 107 leading experts from fifty-two nations of whom more than half were from the developing world. The report called for a comprehensive review of global diet and land use, emphasising the need for a 'co-ordinated response' to shift towards more plant-based diets and sustainable land management systems to protect soil, conserve forests and arrest degradation. 'Land,' the report concluded, 'is part of the solution'.[90]

With their emphasis on the systemic links between diet, culture, health, ecology and climate, such reports are among the most authoritative and comprehensive arguments yet for why business as usual is no longer an option. When it comes to food and farming in the future, nature must come first.

The Supreme Farmer

> The maintenance of the fertility of the soil is the first condition of any
> permanent system of agriculture.
>
> Sir Albert Howard[91]

Apart from being vital to biodiversity, the most obvious benefit of farming with nature is that it allows, indeed requires, us to live closer to
mother earth. Such closeness is essential to organic or natural farming,
not just so as to be able to keep an eye on one's plants and animals, but
because of what Fairlie calls the 'geography of muck': the need to recycle nutrients. If one wants to feed crop residues to Daisy and Buttercup,
they need to be in a nearby field, not off in some far-flung feedlot.[92]
Natural farming is, by definition, mixed and local, since, as the English
agronomist Sir Albert Howard noted in his 1940 book *An Agricultural
Testament*, that is how Nature herself 'farms'.

Howard, the *de facto* father of modern organic farming, called nature
the 'supreme farmer', since, for millennia, she has produced maximum
abundance with minimum waste in every conceivable terrain. The key
to successful farming, he argued, must therefore lie in following nature
as closely as possible:

> Mother earth never attempts to farm without live stock; she always
> raises mixed crops; great pains are taken to preserve the soil and to
> prevent erosion; the mixed vegetable and animal wastes are converted
> into humus; there is no waste, the processes of growth and the pro-
> cesses of decay balance one another; ample provision is made to
> maintain large reserves of fertility; the greatest care is taken to store
> the rainfall; both plants and animals are left to protect themselves
> against disease.[93]

Howard's thinking was remarkable, not least because it flew in the
face of the prevailing agricultural orthodoxy. Ever since Justus Liebig's
discovery that plant growth depended chiefly on nitrogen, phosphorus
and potassium, farmers had used chemicals to improve their crop
yields.[94] For Howard, however, something was missing from this rosy

picture. In 1905, he leaped at the chance of becoming the imperial economic botanist in India, largely because the job came with seventy-five
acres of land on which he could experiment. Observing local farmers,
he found that they grew healthy, abundant crops year after year without
any chemical fertilisers or pesticides. Copying their methods, he soon
found that he too could produce healthy crops without the need for
any external inputs. As he later recalled:

> By 1910 I had learnt how to grow healthy crops, practically free from
> disease, without the slightest help from mycologists, entomologists,
> bacteriologists, agricultural chemists, statisticians, clearing-houses of
> information, artificial manures, spraying machines, insecticides, fungi-
> cides, germicides, and all the other expensive paraphernalia of the
> modern Experimental Station.[95]

Howard had begun to piece together the life cycle that powers our
edible planet and the crucial role played in it by microbes. Without
recourse to modern microscopy, he could only guess at what microbial
trade-offs were taking place beneath his gaze, yet his naturalist's mind
told him that the picture he'd formed was basically correct, an intuition
that has proved remarkably sound.

Living Soil

At the heart of it all, said Howard, is the living soil. The central exchange
of the natural economy, its complex communities use physical, chemical and biological processes to transform dead matter back into living
things. Multiple beings above and within it harness the sunlight, water
and minerals needed to fuel the eternal cycle of growth and decay.
Although plants are vital to the process due to their ability to perform
photosynthesis, animals, insects, worms and microbes play a key role
too, helping to break down organic matter to create humus, the darkish-
brown substance that gives soil its characteristic colour and texture.
Humus is the key to fertility, said Howard, since it is the living medium
through which plants and soil exchange energy and nutrients. This

either happens directly via the plant's roots, or through living bridges called mycorrhizae (from the Greek *mykēs*, fungus + *rhiza*, root) forged between plants and soil fungi. Both types of exchange depend on the crumbly, aerated structure of humus: the first, because root hairs suck mineralised water directly from porous soil walls, and the second, because soil fungi eat humus and then pass on the nutrients directly to plants.

Howard noticed that crops grown in healthy soils had a marked resistance to disease. He noted the steep deterioration in crop resilience when ox ploughs and manure-based compost were replaced by machines and chemicals.[96] 'Machines do not void urine and dung,' Howard observed, 'and so contribute nothing to the maintenance of soil fertility.'[97] Animal waste was crucial to soil and compost, said Howard, since without it diseases quickly spread. Indeed, it was clear that plants preferred 'humus made with animal residues', to the extent that sickly ones could be restored to health with a remedial dose.[98] When Howard returned to England in 1934, he had an ideal opportunity to demonstrate this form of plant medicine, since his garden contained a diseased, pest-ridden apple tree. He lost no time in applying his compost to its roots, and within three years the tree was restored to the full bloom of health, with no pests to be seen. 'We need not strive after quantitative results,' declared Howard, 'the qualitative will often serve.'[99]

Such views naturally put Howard on a collision course with what he called the 'NPK mentality' of 'Liebig and his disciples', who failed to see that the 'many-sided properties' of humus could never be replaced by chemical manure. Forged by living creatures, humus was teeming with 'a vast range of micro-organisms' that formed an 'important section of the farmer's invisible labour force'.[100] Humus, in short, was complex, diverse and alive, while chemical fertilisers were simple, uniform and inert. To expect a sack of chemicals to perform similar miracles to decent compost, said Howard, was to betray 'a fundamental misconception of what soil fertility implies'.[101]

Although Howard gained some influential allies such as Lady Eve Balfour, co-founder of the British Soil Association, his chances of achieving real reform were cut short by the outbreak of war.[102] Maximising short-term yields was the understandable aim of wartime, yet the real damage was done once the war was over, when Allied governments encouraged munitions factories geared up to make poisons and explosives to make pesticides and fertilisers instead. Keen to keep the factories going in case war should break out again, ministers encouraged farmers to keep calm and carry on using chemicals. As far as the agricultural war was concerned, Liebig's disciples had won.

Talking Trees

Today, modern microscopy has confirmed that Howard was not only right about soil, but that he was on to something far bigger (and smaller) than he imagined. Mycorrhizae are now known to occur in some 80 per cent of flowering plants, including all major edible human crops, playing a crucial role in the vitality of both plant and soil.[103] In exchange for the mineral-rich nutrition that plants receive directly from fungi, they provide carbohydrates in return, in the form of sugary exudates secreted from their roots. Mycorrhizae thus represent a nutritional exchange in which plants and fungi trade sugars and minerals to their mutual benefit.

As David R. Montgomery and Anne Biklé explain in *The Hidden Half of Nature*, mycorrhizae form a crucial link in the chain of life, representing the point where geology and biology meet.[104] The ultra-fine hair-like parts of soil fungi known as hyphae (often spotted on upturned rocks) can multiply the effective surface area of a plant's root system tenfold, boosting its uptake of key nutrients such as phosphorus. As well as being highly efficient food hubs, mycorrhizae are also a living inter-species communication system. Like the wires of a telephone exchange, they can stretch for many miles – those in ancient forests are the largest living organisms on earth – carrying signals between plants and soil. Forget the Internet of things; the most complex communicative network on earth is many millennia old and its messengers aren't bits, they're microbes.

We are only now starting to piece together how plants, fungi and humus communicate. The rhizosphere (the zone surrounding a plant's roots, named by the German agronomist Lorenz Hiltner in 1904) is swarming with beneficial microbes, deliberately recruited by the plant with specific exudates. Such microbes act as 'palace guards', protecting their host by identifying potential pathogens and replacing or repelling them as necessary. 'Seducing microbes with sugars,' Montgomery and Biklé note, 'is the heart of the botanical world's defence strategy.'[105]

The value to plants of such partnerships is evident from the huge investment they make in them: exudates can account for up to 40 per cent of a plant's total sugar output.[106] In addition to such sweet treats, plants lavish their microbial partners with a feast of amino acids, vitamins and phytochemicals. As Hiltner guessed, a plant uses its phytochemicals strategically, both to recruit specific microbes and to direct their actions; in return, the recruits send messages to the plant warning it of impending danger and triggering a defensive immune response. Via the mycorrhizal highway, plants under attack can also send warning signals to one another. J. R. R. Tolkien wasn't so fanciful after all: trees really can talk.

Such discoveries challenge the foundational principles of industrial agriculture. To begin with, they show why ploughing – practised by farmers since time immemorial – is so damaging, since it breaks the mycorrhizal networks so vital to both plants and soil. As the Japanese farmer Masanobu Fukuoka showed, however, we can farm perfectly effectively – indeed far more productively – *without* ploughing. Aiming to live as close to nature as possible, Fukuoka developed what he called natural or 'do-nothing' farming, growing alternating crops of rice, rye and barley by sowing the new crop directly over the old one before harvesting, then spreading the straw from the harvested crop on the field to discourage weeds. 'There is probably no simpler method for growing grain,' Fukuoka wrote in his 1975 book *One-Straw Revolution*; 'it involves little more than broadcasting seed and spreading straw, but it has taken me over thirty years to reach this simplicity.'[107] As well as being very little work, Fukuoka found, his method of natural farming (also referred to as 'no-till') was highly productive, consistently attaining yields as high or higher than those on neighbouring farms. 'This method completely contradicts modern agricultural techniques,' he wrote. 'With this kind of farming, which uses no machines, no prepared fertiliser and no chemicals, it is possible to attain a harvest equal to or greater than that on an average Japanese farm.'[108]

As Liebig himself acknowledged towards the end of his life, his assumption that organic matter was irrelevant to soil fertility was disastrously wrong.[109] Although plants do need plenty of nitrogen, potassium and phosphorus in order to thrive, feeding them doses of NPK disrupts the natural balance between plant and soil. Indeed, plants given regular shots of free chemical nutrients switch off their flow of exudates, no longer bothering to recruit microbial helpers to their aid. The result is a depleted rhizosphere that deprives plants of micronutrients and weakens their resistance to disease. Dosing crops with chemicals is like feeding your kids on fast food: the children may grow rapidly (mostly sideways), yet their overall health will suffer.

The analogy is more apt than it may appear. Since plants form the basis of all our food, narrowing their diet and weakening their immune systems has a direct impact, not just on their health, but on ours. Modern microbiology has revealed the direct relationship between plant and human health: in many ways, the rhizosphere and the gut are directly analogous. Both are by far the most complex part of their respective hosts' microbiomes, and both depend on a rich and diverse community of microbes (some of which are in common) in order to function. The reason why silver-bullet solutions will never work when it comes to feeding ourselves is because complexity, for all living things, is the key to health.

Embracing Complexity

> The health of man, beast, plant and soil is one indivisible whole.
>
> Lady Eve Balfour[110]

It is somehow fitting that our habit of peering down microscopes is revealing the extent to which our past attempts to manipulate nature have been misguided. Microbiology is a field capable of healing our bodies and landscapes as well as our compartmentalised thinking. Our rediscovery of what the ancients knew – that our health is intimately bound up with the natural world – has the power, if we let it, to transform the way we eat and live.

Instead of fighting an unwinnable war against pathogens and pests, we could instead be recruiting armies of friendly microbes ready and willing to fight our battles for us. Our fear of microbes is another result of tunnel vision, fuelled by the unfortunate fact that the small minority of microbes that are major pathogens happen to be those that can be easily cultured in labs. The 'germ theory' of disease (based on Louis Pasteur's discovery in 1859 that germs were present in air and the identification by Robert Koch of those responsible for anthrax in 1876) has understandably focused on combating such lethal pathogens. It has given us vaccination, sterilisation and antibiotics, without which many of us wouldn't be here, but its negative legacy has been to leave us with a distorted and fearful view of the microbial world.

Ironically, it was Pasteur himself who made the crucial breakthrough in our understanding of the positive benefits of microbes, with his discovery in 1857 of the role of yeast in fermentation. Crucial to the manufacture of many of the foods without which life would be barely worth living (bread, cheese, yoghurt, wine, beer, coffee and chocolate, and, for those who like it sour, kimchi and sauerkraut), fermentation relies on microbial action to transform sugar into acids, gases or alcohol. The result, as our Neolithic ancestors discovered, no doubt to their delight, not only produces an array of delicious and complex flavours, but preserves fresh foods, sometimes for years, into the bargain. Given that Pasteur discovered the miracle-workers behind this time-honoured process, it is a further irony that, having identified the 'bad' bacteria that soured milk, he invented a way of preserving the liquid by killing *all* its microbial life, the heating process we know today as pasteurisation.

If you're thinking that pasteurisation has something in common with pesticides, you would be right. Blanket solutions to complex problems often create more dilemmas than they solve. That is not to say that they are without value: none of us would want to live in a world without antibiotics, although, thanks to our demand for cheap meat, that now seems a distinct possibility. But by treating the living world too simplistically, we run the risk of preventing natural systems from doing an often far better job.

Take, for example, the case of Salers, a hard cheese made high in the French Auvergne, so ancient and noble that it even gets a mention in Pliny. Only five families still make the cheese in the traditional way and, when you find out what's involved, you won't ask why. Milking their notoriously cranky cows out in the field twice a day, the farmers must tether the calves to their mothers' legs to persuade the latter to give up their milk, which amounts to just one third the volume obtained from a docile Holstein. The milk is then taken back to the farmers' huts, where it is poured into wooden vats called *gerles* which are used for years without disinfectant and are thus teeming with lively microbes. Once microbes from both *gerles* and milk get to work fermenting the latter's sugars, they transform the liquid into a unique, pungent cheese.

Rare and delicious, Salers Tradition is in high demand. Yet as Bronwen and Francis Percival related in *Reinventing the Wheel*, its 2,000-year tradition nearly came to an abrupt end in 2004, when the French food safety agency decided that raw milk souring in dirty wooden tubs was a disaster waiting to happen. The cheese was about to be condemned when a saviour emerged in the form of microbiologist Dr Marie-Christine Montel, director of the Institut National de la Research Agronomique (INRA). Hauling some *gerles* off to her lab, Montel found their microbial communities to be so lively that raw milk was inoculated within seconds of making contact. Not only that, but any contamination with pathogens, she found, was actively and robustly resisted. The old wooden *gerles*, in short, were naturally stable, disease-resistant, perfect starter factories. Raw milk was also vital to their function, Montel discovered: when she introduced pasteurised milk, the microbial communities became unbalanced and ineffective. With Montel's official blessing, Salers Tradition lives on.

The idea that microbes can be positively beneficial, even protective, is transforming our view of the natural world. Such knowledge has of course been around unofficially for decades: as the daughter of a doctor and a nurse, I was always encouraged to eat food that had fallen on the floor, on the basis that this would strengthen my resistance to germs. I have done this all my life and have one of the strongest constitutions I know, yet have mostly kept the advice to myself, on the basis it might sound a little weird. Today, however, my parents' views are being backed up by modern science. One recent study, for example, found that children brought up with pets or close to farms were far less likely to suffer from allergies than those brought up in immaculate, super-clean homes.[111] Indeed, the recent rise in the West of allergies like asthma is thought to be linked to our obsession with hygiene. Like a healthy *gerle* or rhizosphere, our microbiomes – the core of our immune systems – need to be fed a rich and varied diet in order to learn their trade. Our guts, like our brains, need to be educated.

Microbial Masterminds

The more you find out about microbes, the clearer it gets why our Western diet is killing us. Nutrition is a relatively new field, which has, as we saw earlier, been dominated for the past century or so by celebrity gurus, quacks, cranks and the food industry. Today, as modern microscopy lifts the lid on what is really going on in our guts when we neck a Coke or devour a doughnut, it's time to fight back. Our days as credulous victims of the fad-peddling food and diet industry are over.

The reason why complexity matters in food – both in its variety and its molecular structure – is because we eat to feed not only ourselves, but the roughly 3,500 microbial species that live in our gut and keep us healthy and happy. Only recently has the pivotal role played by such microbes become apparent, not just in moderating our moods and behaviour, but in protecting us against the diseases of affluence – obesity, diabetes, cancer, heart disease, bowel disease and dementia – that have become our biggest killers. The fact that we are suffering such a crisis is almost entirely down to our dodgy diet: too little variety, too many refined sugars and, crucially, not enough fibre. Complex carbohydrates such as cellulose – happily digested by cows – pass straight through our upper digestive tract, thus making a fabulous feast for the microbes in our colon, which constitute three quarters of our microbiome and 80 per cent of our immune system.[112] Fibre, in short, is the rations we feed our microbial armies, just as plants ply theirs with sugar.

Kellogg and Graham may have been cranks, but they were on to something: a diet of rich, refined, sugary foods really is bad for us.[113]

Although less preachy than his American forerunners, the British genetic epidemiologist Tim Spector is equally firm in this view: there is, he says, a 'straightforward link' between microbial diversity and health.[114] A leading expert on the human microbiome, Spector knows a thing or two about guts. In 2016, he travelled to Tanzania with the BBC's *Food Programme* to spend three days living and eating with the Hadza, among the last hunter-gatherers on earth.[115] With a diet consisting of milky, tangy baobab juice for breakfast, followed by snacks of berries, tubers, honey and grubs with the occasional piece of porcupine, the Hadza are what Spector calls 'microbiome superstars', with some 40 per cent more microbial diversity than the average Westerner. Their microbes also include many rare species unknown in Europe, some of which could be crucial in helping us to stay slim or fight disease. The race is therefore on to discover such microbes before they are lost for ever.

When he returned to London and analysed stool samples he'd taken while staying with the Hadza, what Spector found astonished even him. In the space of just a few days, his microbiome had changed beyond recognition: there was far more diversity, with dramatically heightened levels of microbes associated with slimness such as *Akkermansia* and *Christensenella*, and of *Faecalibacterium prausnitzii*, which dampens inflammation. Fully 2 per cent of Spector's microbes now consisted of an entirely new group unknown in the West, yet known to be present in plants and soil. Such microbes could have amazing health properties, says Spector; we simply don't know yet. Nevertheless, he adds, 'If I had a yoghurt that did that for me every day, I'd certainly buy tons of it.'[116]

Creatures of the Wild

> There is a general duty of humanity, that attaches us not only to animals, who have life and feeling, but even to trees and plants.
>
> Michel de Montaigne[117]

Spector's research gives us some idea of what 12,000 years of civilisation have cost us. Living and eating in the wild, the Hadza are about as in tune with nature without and within as it's possible for humans to be.

They epitomise the ideal balance that modern life makes all but impossible: they have no cars or computers, yet neither do they have cancer or heart disease. Whether or not the swap has been worth it is moot; it is simply the result of the evolutionary path we humans have taken.

Wherever and however we live, we remain living mirrors of the natural world – or rather, of the modified versions that we ourselves have created. Only now, as foraging people like the Hadza stand on the brink of extinction, are we starting to grasp what that means. By simplifying nature, we have diminished ourselves. *Gilgamesh's* authors were right: for us humans, living in cities away from the wild is a form of death.

All is not lost, however, as Spector's miraculous microbial makeover proves. Indeed, the remarkable speed with which ancient microbes seem willing to colonise urban humans heralds a potentially thrilling new chapter in our relationship with nature. In some senses, our technological journey has brought us full circle, reminding us of what our forager forebears instinctively knew: that only when nature flourishes in all its fullness, do we in ours.

The implications of this rediscovery are profound, not just for the way that we eat and live, but for the philosophical basis of civilisation itself. At whatever scale you look at it, our task at this pivotal moment in our evolution is not just to reconcile ourselves with nature or to save the world by carving it up like some colossal pie, but to recognise ourselves as creatures of the wild. In an age of rampant urbanisation and technical mastery, this may seem a strange way to describe ourselves, yet that is the whole point. What our reawakened sense of deep connection with nature shows us, above all, is just how deadly the deal we have struck with it really is. If we are to have any chance of thriving in the future, we need to recalibrate that deal, and fast.

How then might we eat, live and think like creatures of the wild? Most obviously, it means respecting and preserving the world's great wildernesses, which in turn means knowing what *not* to eat. Obliterating rainforest for palm oil or bottom-trawling the seabed for fish (the underwater equivalent of dragging a 30-tonne, 150-metre iron bar across the countryside) are simply not the behaviours of civilised beings. Instead, we must learn to eat with a lighter touch, in ways that preserve wilderness, rather than destroying it. We have plenty of evidence of how our ancestors did it and of how some humans still do, but could such practices translate into a modern urban existence? The surprising answer is: yes they could, more easily than you might think. Take food, for example. We already know that wild foods are vastly more nutritious than cultivated ones. The wild berries the Hadza eat, for instance, have between ten and a hundred times the nutritional content of blueberries bought in Aldi or Asda. Instead of breeding such wildness out of plants in order to achieve higher yields, therefore, we might start farming in such a way as to preserve their wildness. That is, of course, pretty much what organic farming already does. But we could go much further, by mimicking and encouraging wild growth to produce edible ecosystems almost as rich and diverse as those in the wild. Such an approach would clearly require us to eat very different foods (less bread and pasta and more nuts and berries, as well as more seasonal produce), yet they would also be deeply efficient in the sense that they would focus on quality rather than quantity. If we ate berries a hundred times more nutritious than supermarket ones, we'd only need to grow one hundredth the amount.

Eating like creatures of the wild would also require us to expand our diet to include far more species, including insects. Two billion people worldwide already dine on the creatures we call creepy-crawlies; as creatures of the wild, we too might learn to love their crunchy umami taste and to value this nutritious, low-impact source of protein. We should also re-embrace some bitterness in our diet. As the medical herbalist Alex Laird points out in *Root to Stem*, the bitterness we taste in foods such as fruit pith, pips and skin is a sign of the presence of phytochemicals, the natural chemicals in a plant's defence armoury and healing mechanisms.[118] Most drugs and medicines, lest we forget, originally came from plants, and work because they irritate – and thus stimulate – our immune systems enough to keep them ready for action, just as our muscles need regular exercise in order to keep us fit. Bitter foods such as cabbage and chicory – which the food industry has been busily making sweeter to appeal to our taste buds – are thus to be savoured. In food, as in life, we need friction in order to flourish. Next time you hesitate over the Brussels sprouts, therefore, spare a thought for the wild companions in your gut: it's not just you who likes to feast at Christmas. If you cater for your microbial companions whenever you plan a meal, you'll eat much better – and you'll never dine alone.

Eating Nature

For the US chef Dan Barber – whose signature dishes at his renowned restaurant Blue Hill at Stone Barns include a home-grown carrot steak and baby vegetables on a rack – the secret lies in understanding the deep connection between natural complexity and deliciousness. The revelation came when Barber was sent a shrivelled corncob in the post from heritage seed collector Glenn Roberts, accompanied by a $1,000 cheque and a note inviting him to grow it. Sceptical, Barber asked his vegetable grower Jack Algiere to raise the cob, which he did using the Three Sisters method traditional to the Iroquois, planting it alongside dry beans and squash (the corn stalks provide support for the beans, which in turn supply nitrogen for the corn, while the ground-hugging squash deters weeds). Several months later, when Barber made polenta with the resultant corn, it was nothing short of an epiphany. 'It wasn't just the best polenta of my life,' he wrote in *The Third Plate*, 'it was polenta I hadn't imagined possible, so *corny* . . . the taste didn't so much disappear as slowly, begrudgingly fade. It was an awakening.'[119]

What Barber realised that day was that the way in which chefs usually work – writing their menus and then finding producers to supply the ingredients – had to be turned on its head. Instead of deciding what to eat and then expecting nature to supply it, we need to ask the landscape what it wants to grow. The key to deliciousness, he understood, lay not in the kitchen, but in the complexity of nature. 'A bowl of polenta that warms your senses and lingers in your memory,' he wrote, '. . . speaks to something beyond the crop, the cook, or the farmer – to the entirety of the landscape, and how it fits together.'[120] Today, Barber is a global advocate, not just for nature-led eating, but for a return to the traditional complexity and zero-waste mentality that was the basis of all the world's great cuisines. His latest venture is a seed company called Row 7, which combines the breeding of heritage seeds with cutting-edge (non-GM) technology to bring the very best natural flavours to ordinary cooks. For Barber, the key to eating well is to respect nature. As he puts it, 'good farming and delicious food are inseparable'.[121]

Eating more from nature will clearly have a radical effect on our landscapes, as Knepp Castle Estate, a 3,500-acre aristocratic swathe of West Sussex, suggests. Intensively farmed for decades, Knepp's clay soils were so depleted by 2000 that the estate was on the verge of bankruptcy. The owners, Charlie Burrell and Isabella Tree, decided to reverse the damage by doing what seemed most obvious: letting nature back in. They stopped farming grain (always marginal on such heavy soils) and instead started to rewild the farm, allowing the natural habitats, including scrubland and wetlands, to restore themselves: 'Rewilding,' Tree says, 'is restoration by letting go.'[122] Today, Knepp is a rich, diverse, yet still productive landscape in which longhorn cattle, red deer, wild ponies and Tamworth pigs roam freely in a habitat that is rapidly recovering its former biodiversity, with a wide range of plants re-establishing themselves and a profusion of insects, bats and birds, including nationally threatened turtle doves and nightingales. Although Knepp no longer feeds as many people as it once did, it demonstrates that productiveness and wildness are not mutually exclusive.

Is it possible to transform our cities in the same way? Most cities already have parks, gardens, ponds and rivers that could easily be subjected to the Knepp treatment, as well as a host of wild and semi-wild beings – plants, birds, insects, hedgehogs, foxes, rats and mice – that we already live alongside. Beyond making such spaces wilder and more productive, what else might we change? The key shift would arguably take place in our heads, so that we once again recognise that our urban lives depend on wild ecosystems, without which we would rapidly cease to exist. Calling ourselves creatures of the wild is one way of shedding our habitual – and deadly – anthropocentrism.

Thinking on its own isn't enough, but it is a vital first step. Merging back into the wild, towards conscious coexistence with our fellow non-humans, farming in ways that mimic and complement nature, rewilding cities and farmland – these are among the acts that will help us find our way home. We need biodiversity, not just because it's necessary to our survival, but because we're part of it. We carry wildness within us, we can't escape or transcend it, it is the daemon that we have lost and yet yearn to bond with again.[123] Rethinking the way we live is a long-overdue exercise; yet more urgent still is that of rethinking ourselves.

Emerson was right: we feel at home among trees because we co-evolved with plants: trees are our family, and we theirs. The Japanese practice of 'forest bathing', of spending time immersed in the forest, has been shown to reduce stress and anxiety, to increase energy and boost the immune system.[124] Practised by one quarter of all Japanese, such immersion in nature, even for short periods, is increasingly recognised as a natural remedy for overstressed, urbanised humans.[125] Reconnecting with wildness, even if only via a walk in the park, is a vital post-Cartesian corrective, reminding us that no animal – even a technically enhanced human – is a machine.

Law of Return

I am deep in a forest-like garden. The foliage is dense, yet there are clearings ahead; I can just about see a way forward. Up above, tall alders spread their filigree canopy, casting dappled light on the shrubs and bushes below. Some of the plants here are exotic – one with vast spreading leaves looks like dinosaur salad – yet some are familiar: I'm pretty sure the bush I've just passed was a wild raspberry. I plunge further in and unexpectedly come into a glade, with some makeshift benches under a tree and a large lily pond that Monet would no doubt have loved to paint. The garden feels unkempt, yet there is a gentle sense of order to it, not unlike how one might imagine the Garden of Eden.

This is Dartington Forest Garden in Devon, designed and planted in 1994 by British agro-forestry expert Martin Crawford and cared for by him ever since. This two-acre plot may well be the farm of the future. Although not specifically designed to maximise productivity, its low-maintenance, low-input, year-round fecundity is, Crawford believes, the key to how we might one day feed ourselves. Most of the plants here are edible, but by no means all: some have medicinal properties and others provide specific ecosystem services, for example providing ground cover to discourage weeds or aromatics to deter pests.

The key to forest gardening, Crawford explains, is to create ecosystems that, like those in the wild, sustain a natural balance between inputs and outputs. The alders, for example, were pioneer trees, planted early for their ability to fix nitrogen, for which they remain the garden's main source. Valerian, meanwhile, is grown as a mineral accumulator, its deep roots absorbing nutrients that other plants can't reach. Once it has flowered, it is cut back to release those nutrients into the soil for other plants to use.

Like all plants in the garden, the food crops here are all perennials. Alongside familiar varieties of fruits, nuts and berries such as apricots, chestnuts and currants, is an array of some 140 species, including white mulberry (a tree vegetable like spinach), ostrich fern (with asparagus-like shoots), sweet cicely (for its edible roots and fennel-tasting leaves), bamboo (prized for both its shafts and its edible shoots), Babbington's leek (a perennial version of the Welsh national emblem) and Turkish rhubarb, the aforementioned dinosaur salad that turns out to taste like gooseberry. Since all the plants are perennials, Crawford explains, there is no need to plough or disturb the soil, allowing soil fungi – 'the most important organism of all' – to flourish.

The garden has no problem with slugs and snails, Crawford continues, since the frogs and ground beetles eat them, and there are very few weeds, since he keeps the earth covered with ground-hugging plants such as false strawberry, which the beetles also love. Crawford does, however, have a squirrel problem, due to the fact that the garden is right next to a pine wood. For this reason, he doesn't grow as many nuts as he'd like, and when his chestnuts fall to the ground, he has to race the squirrels to get them.

Such bushy-tailed aggravations apart, looking after the garden is easy, says Crawford. He prunes his trees so that he can reach their fruit more easily and weeds once a month between April and July, a four-hour task that involves chopping back certain plants in order to give others a competitive advantage. Now that the original work of designing and planting the garden is over, he says, his approach is largely *laissez-faire*: he believes in letting plants do what they want, since they are far happier and healthier that way. Wild raspberries, for example, like to move: if they can't, they become stressed and prone to disease. The ones we passed earlier, he tells us, were planted twelve metres from where they are now. It's largely a case, says Crawford, of letting natural systems work for you, with just the occasional tweak, much as our forebears no doubt did when they started shaping wild forest. As well as being light, Crawford adds, the work is always interesting; it certainly beats the 'tedious hours spent hoeing rows of carrots' from his market-gardening days.

Could forest gardens really be the farms of the future? In our world of shrinking resources and increasingly extreme weather, they've certainly got a lot going for them: minimal input, maximum nutrient and water retention and natural resilience due to their great diversity, which allows for constant adaptation as climatic conditions shift. Like no-till farming and wild farms such as Knepp, forest gardens maintain what Albert Howard called the Law of Return: the natural balance between growth and decay essential to all fertility. With the addition of poultry and pigs, which naturally forage in forests, the garden's productivity could be further increased to provide a naturally rounded, semi-wild diet.

With forest gardens and wild farms on our doorsteps – perhaps in rewilded parks and gardens as well as in the countryside – we could all live closer to nature. With minimal upkeep and something new to be harvested each day, such farms and gardens could become our new commons. We could visit them as often as we chose, to forest bathe, do some planting or pruning, gather something for dinner or just hang out. We would be physically fitter, soothed, entertained and engaged in a productive, meaningful common activity.

Star Dust

Is such a vision fantasy – just an echo of our yearning to get back to the Garden, to paraphrase Joni Mitchell? Forest gardens aren't the whole answer, of course, any more than vertical farms, vegan diets or garden cities, yet they could certainly be a useful piece of the puzzle. One could argue that forest gardens are true vertical farms, contrasting our two-dimensional, monocultural agri-food model with three-dimensional, complex fecundity.

However we live in the future, embracing complexity will be key to our flourishing. Joseph Hayek once said that markets were too complex for anyone to understand and that they should therefore be allowed to run free. How ironic then that when it comes to nature – an infinitely more complex system – his disciples argue for its control and containment. Could it be that they see profit to be made in letting money run free, but not nature?

Robots are already farming our land; before long we may create cereals that can fix nitrogen. We can't stop the march of progress, but we can decide who owns such technologies and how and when we use them to help us balance our lives with nature. Non-invasive, commonly owned, technically enhanced natural farming is probably our best hope of doing just that. As long as we let the Supreme Farmer be our teacher, we can't go too far wrong. However smart we think we are, we would do well to remember that the plants to which we owe our lives have had a 700-million-year head start.

Part of our unwillingness to admit our complicity in the natural order is fear of acknowledging our part in the eternal cycle of growth and decay. Denial of death, as we saw earlier, is a peculiarly Western affliction, a result of our long struggle to subjugate nature and thus free ourselves from her mortal grasp. The most precious gift that our technical vision may yet give us is the revelation, played out in the microscopic theatre of truth, that it's time we gave up the fight. Arguably the greatest insight yet on our human voyage of discovery is that we've been inextricably part of nature – and thus staring back at ourselves – all along.

7
Time

A Garden Shed

It's late October, and I'm sitting in a shed at the bottom of one of those unfeasibly large, wooded back gardens you get in certain parts of Victorian north London. The last of the afternoon sun is just catching the leaves, turning them red and gold. It's a very welcoming shed: painted a tasteful shade of pale green, it has large windows, warm lighting and is big enough to seat a dozen of us comfortably around a long table and on a sofa covered with rugs and scatter cushions. The table has a gingham cloth and lots of candles and there is a promising-looking sideboard laden with tea and cake. I am glad of all these comforting touches, since I am here to talk about our last great taboo: death.

This is my first Death Café, an informal gathering of strangers who come together over tea and cake to discuss mortality. The idea was originally dreamed up by the Swiss sociologist Bernard Crettaz, who realised that his students had a great unmet need to talk freely about death. If he could bring people together in a reassuring setting such as a restaurant or café, Crettaz thought, then the barriers might come down. He was right. The Café Mortel, first held in a Paris bistro in 2004, attracted 250 people of all ages and was soon a regular event.[1]

The main aim of the Café Mortel, said Crettaz, was to listen. There were just two house rules: one had to speak honestly and refrain from preaching. The café ran successfully for several years until Crettaz decided that he'd quite like to get back to his day job, at which point the concept itself might have died, had not a London web designer by the name of Jon Underwood happened to read an interview with Crettaz and realise that he had found his life's vocation.[2] Rebranded Death Café by Underwood in 2011, by 2019 the movement had held more than 8,000 meetings in sixty-five countries, although Underwood didn't live to see it, having died suddenly in 2017 at the age of just forty-four.[3]

Back in the shed, having helped ourselves to provisions, we sit in some trepidation waiting for proceedings to start. We are a mixed bunch, rather younger than I expected, with several people in their early thirties and the oldest among us a woman I reckon to be in her mid-seventies. For some reason, I had expected there to be more women than men, but in fact we are more or less evenly split. Our host for the evening is Gemma, a thirty-something celebrant whose encouraging smile and warm tone help to reassure us. There is no agenda this evening, she says, apart from an initial round of introductions, after which we are free to take the conversation anywhere we please.

After an awkward pause, a young Indian doctor tells us that he is here because he works in palliative care and finds there aren't enough opportunities to talk about death at the hospital. He finds Death Cafés really helpful, he adds – this is his sixth. Next, a young Polish designer explains that he has come because it is close to All Souls' Day in Poland – a festival during which families gather to honour their dead – and he is keen to see what we get up to in Britain. (I resist telling him that we generally dress up as witches and ghouls and bully strangers for sweets.) Next, a worried-looking middle-aged man says he is here because he doesn't believe in God, yet is keen to explore whether there might nevertheless be an afterlife. The older woman says that she used to believe in God, but doesn't any more, and wonders what all that was about. A Christian minister looks tempted to respond, but confines himself to informing us that he has come on a recce because he plans to host a Death Café of his own. A young woman then explains she is here because a friend recently took her own life and she is thinking of joining the Samaritans. So it goes on round the table. Our reasons for being here are as varied as our age, gender, backgrounds and beliefs: death, as they say, unites us all.

After the initial hesitancy, the conversation is soon flowing freely. We discuss our fear of death, from which, perhaps unsurprisingly, the septuagenarian seems to suffer the most, and then move on to the afterlife, during which the minister threatens to transgress Crettaz's golden rule by coaxing the old lady back into the flock. Intrigued by this evolving subplot, I wonder aloud how having a faith in God might affect one's natural fear of dying. A cancer nurse next to me, who has clearly witnessed her fair share of death, says that, for her, believing in God doesn't really help: she just likes to think there is a better world waiting for us somewhere than the messed-up one we live in.

We move on to other topics, discussing euthanasia, living wills and how one might best plan for a 'good' death and funeral. Apart from the matter of whether there is an afterlife, where we seem evenly split down the middle, we agree on most things, especially on the need to break down the taboos surrounding death. Just as we are getting really stuck in, Gemma tells us that our time is nearly up, although she says we are welcome to stay on and finish the tea and cake. I am amazed: two hours talking to complete strangers about mortality have absolutely flown by. I feel I could carry on for hours. I think I may be on the verge of becoming a Death Café habitué.

The Last Taboo

> Do not go gentle into that good night. / Rage, rage against the dying
> of the light.
>
> Dylan Thomas[4]

Why are most of us so bad at talking about death? And what, for that matter, do we consider a good death to be? The questions are related and both depend partly on whether or not one believes in an afterlife. On that matter, our group turned out to be remarkably representative: a 2017 survey carried out for the BBC found that identical proportions of Britons – 46 per cent – do and don't believe in an afterlife, with 8 per cent remaining unsure.[5]

For a secular nation like ours, such figures may seem surprising. Compared to the US, where 'In God We Trust' is inscribed on every banknote and politicians thank God after almost every speech, religion plays a relatively minor role in British public life. Yet whatever one's view of a life hereafter (in which 80 per cent of Americans believe), something both nations share is a widespread denial of death.[6] Belief in an afterlife doesn't, it seems, necessarily cure one of the fear of dying; on the contrary, as the cancer nurse in the shed assured me, the two can quite happily coexist. Even those who look forward to a celestial life hereafter think it will be very different to the terrestrial one, and the transition from one to the other can hold just as much dread.

Whatever your view of the afterlife, facing death is undeniably hard, but that still doesn't explain why we have such difficulty dealing with it. Part of the problem is arguably down to the fact that dying these days is so medicalised. We have become remarkably adept, if not at cheating death, then at least at postponing it: diseases that would have carried off our forebears in days can now be resisted for months or years as we 'battle' whatever it is that will do for us in the end. In our post-industrial world, the desire to live as long as possible is understandable. Most of us lead lives more diverting and comfortable than any humans in history; we'd hardly *be* human if we didn't want to stick around to enjoy it all a bit more.

Modern medicine has brought us great benefits yet, as the US surgeon Atul Gawande argues in *Being Mortal*, paradoxically it often robs us of a good death. In the modern world, most of us take our last breath stuck full of tubes in some bleeping, neon-lit hospital ward, not peacefully at home, surrounded by friends and family. With medical training focused entirely on saving life rather than dealing with its end, says Gawande, both doctors and patients often make end-of-life decisions that are unrealistically skewed.[7] Many doctors seek to prolong their patients' lives at any cost, seeing the death of even terminally ill patients as some kind of failure. The result is that patients are put through gruelling treatments that do little more than prolong their suffering, giving them 'a substantially worse quality of life in their last week than those who received no such interventions'.[8]

We forget that the dying may have priorities other than simply prolonging their lives. Rather than submit to endless procedures, they might prefer to use their last days going on some long-planned trip or just spending time at home with their family. The choice, in the end, comes down to one of quality versus quantity: would you choose more life at any cost, or swap some longevity in exchange for a better finale? It's a choice that few patients get to make, argues Gawande, since by the time they come to face it, many are too ill to judge, making treatment the default choice. 'We have reached the point,' he says, 'of actively inflicting harm on patients rather than confronting the subject of mortality.'[9]

In order to have better deaths, we need to face up to mortality *in advance*. We need palliative care, not just in the last weeks of life, but as soon as we are diagnosed. One US programme called Coping with Cancer found that, when they were given the necessary support to consider their options, many patients chose to let death run its course rather than fight it, giving up treatments and entering a hospice earlier. As a result, says Gawande, they 'suffered less, were physically more capable and better able for long periods to interact with others. In addition, six months after these patients died, their family members were markedly less likely to experience persistent major depression.'[10]

When it comes to dying well, it seems, acceptance is key. We spare ourselves and our loved ones unnecessary suffering and we are free to enjoy more of what life remains to us, which may, as one 2010 Massachusetts General Hospital study suggests, end up being more than we think. The study found that patients who received palliative care early in their treatment not only tended to suffer less at the end of their lives, but also lived on average 25 per cent longer. As Gawande remarks, such a finding seems 'almost Zen: you live longer only when you stop trying to live longer'.[11]

Living Forever

Such an approach to death is unlikely to go mainstream any time soon. Living longer and staying young are dominant obsessions in the West, with newspapers full of sky-diving grannies and smooth-skinned celebrities whose faces are literally frozen in time. So prevalent is our desperation not to age that it has spawned an entire industry selling everything from superfoods and supplements to injections and surgery, all tagged with the promise of eternal youth.

Beneath all the frozen foreheads and airbrushed selfies lurks the eternal question: if we were to really go for it, how long could we hope to live? So far the evidence is that there are natural limits to how long the human body can last. The oldest person to date, a Frenchwoman by the name of Jeanne Calment, died in 1997 at the age of 122, and the nine runners-up (also women) lived between 116 and 119 years. Although the numbers of centenarians are rising around the world (those in the UK qualifying for a telegram from the Queen doubled from 2002 to 2017), to live much beyond a hundred remains a rare human feat.[12] Despite this, a growing number of people (based mostly in the US) believe that, with the right lifestyle and diet, we can live to 125 and more.

Paul McGlothin and Meredith Averill are at the forefront of this movement. Co-founders of the self-explanatorily named Calorie Restriction Society, McGlothin and Averill argue that, by following a highly restricted dietary regime, we can slow the ageing process and thus greatly prolong our lives. Based at their woodland Longevity Center in Westchester County, New York State, McGlothin and Averill follow a regime punishing enough to make even a supermodel blanch, surviving on 30 per cent fewer calories than the recommended daily intake, consumed in the form of two meagre, mostly vegan meals per day. When British food critic Giles Coren went to film the pair in 2015, he was given what he described as 'punitively coarse' wheat-free bread drizzled in lemon juice for breakfast, followed by a bowl of stewed barley, onions and strawberries topped off with a whole sliced lemon. 'We're not here to show you that we've got the greatest recipes on earth,' McGlothin explained to a horrified Coren, 'but to give you the principles.'[13]

Those principles involve lowering one's blood glucose (which McGlothin measures religiously before and after meals) sufficiently to activate what he calls the body's longevity biochemistry – fasting mode – in which cells shift from growth to maintenance. As McGlothin and Averill explain in their 2008 book *The CR Way*, they were inspired to take up this gruelling regime after reading an article by the geriatrics nutritionist Dr Rick Weinruch, in which he reported that mice fed on a calorie-restricted diet lived on average 34 per cent longer than their well-fed companions.[14] McGlothin and Averill took the radical step of offering themselves as human guinea pigs, to see whether such dietary restrictions could also work on humans. Averill, who favours wrap-around sunglasses and barely speaks in public, admitted in a rare interview to having been planning her 125th birthday 'for absolutely decades now'. While Averill's shades make it hard to assess how she looks on her radical regime, McGlothin has the gaunt features of a man considerably more advanced in years than he actually is. As the TV critic Andrew Billen put it somewhat cattily, if accurately, 'McGlothin and Averill look incredible for a couple in their nineties. Unfortunately, they are in their sixties.'[15]

Looks aren't everything, of course, and McGlothin, at least, gives the impression of a man in love with life. Surviving on gruel, constantly monitoring one's glucose and going for daily walks in a fifteen-kilo body vest (to trick one's bones into staying strong) may not be everyone's cup of tea, yet he seems to thrive on it. 'I've never wanted to part with anything,' he explained to Coren. 'I'm not willing to accept that my eyesight is going to get worse, that my brain's going to get worse, I'm not willing to accept a wrinkle here if there's some way to stop it.' But, Coren objected, isn't that just part of the human condition? 'But *why*?' McGlothin insisted. 'Aren't you special? Is there a time when that specialness should end, that you should become *dust*? I'm not willing to accept that. I'm still a little boy – I've got a long, long time to go.'

If McGlothin and Averill reach their target, I shan't be around to see it, which is absolutely fine by me. Belonging in a certain time feels as important to me as having a place to call home. Just as I have no desire to colonise Mars, I have absolutely no wish to outlive my friends and family. To carry on as some ageing crone, droning on about the days before computers to a bunch of distracted future humans holds about as much appeal to me as living out the rest of my days in a state of semi-starvation. Indeed, it reminds me of one of our favourite family jokes. A man goes to his doctor and asks him how he can live to be a hundred. 'Well,' says the doctor, 'if you give up drinking, smoking and sex, you should have a fair chance.' 'But what if I give up all those things and *still* don't live to a hundred?' asks the man. 'Don't worry,' says the doctor. 'Even if you don't live to a hundred, it'll feel as if you have.'

Building Walls

The paradox at the heart of our relationship with time is that, when it comes to life, more isn't necessarily better. The fatal allure of immortality is a common trope of fables and fairy tales and it doesn't tend to end well. In fact, the general consensus is that immortality is a curse: a life of tedious repetition, apathy and meaninglessness that one yearns to end in death.[16] The ancients had their reasons for insisting that immortality was reserved for the gods. As Uta-napishti tells Gilgamesh when he seeks the secret of eternal life, man was made mortal and must learn to accept his fate. 'You toiled away and what did you achieve?' the sage asks Gilgamesh. 'You exhaust yourself with ceaseless toil, you fill your sinews with sorrow, bringing forward the end of your days.'[17] Gilgamesh is devastated, but is later comforted, as we saw, when he realises that the walls of Uruk will outlast him.

The longing to extend one's physical presence on earth has been a universal human trait. By building walls and temples out of stone, we hope to gain a foothold in time beyond our own fleeting existence. It's only recently, however, that we've tried to cheat death by manipulating our own bodies, an approach that looks set to take off with the rapid advance of biotech. Apart from the fact that it doesn't produce any great architecture, the quest to stay young feels just a tad *immature*: a refusal to grow up that equates to a rather one-dimensional view of life.

One might also argue that it is out of synch with the stage our civilisation has reached. Regardless of what technical innovations may lie ahead, we are clearly in the final stages of the thrusting epoch that gave us the Anthropocene. If the boundless energy of the Industrial Revolution was our cultural adolescence and the twentieth century's feathering of nests a sort of middle age, then our ecologically depleted era is surely analogous to advancing years. Instead of trying to regain our youth, therefore, we might do better learning how to grow old gracefully as a society, living at a slower pace and cherishing the pleasures of such a life.

Some parts of the world are of course at a very different stage in their demographic trajectory. While advanced economies in Europe and Japan are rapidly ageing thanks to lower mortality and fertility rates, others – notably in Africa – are at the start of their transition. Europe reached demographic maturity (with more people over the age of sixty than under fifteen) at the turn of the millennium; by 2050 the world population is expected to reach the same point, with two billion of each.[18] Between now and then, however, both groups will face severe challenges. Ageing nations will face a lack of workers and the burden of looking after their elderly: in Japan, robots have already been recruited as primary school teachers, hotel receptionists and as companions in old people's homes.[19]

Africa, meanwhile, faces the opposite problem, with a population projected to more than double to 2.5 billion by mid-century, by which time 32.2 per cent will be under the age of fifteen.[20] This raises huge questions about how to provide the means for so many to lead a good life. The issue is already pressing in India, where millions of villagers are getting educated for the first time. When the Indian state railways advertised 63,000 jobs in 2019, 19 million people applied.[21] While urbanisation, education and health provision will clearly be key to the future of both India and Africa, the issue remains as to whether such places can evolve without experiencing the pitfalls that industrial nations have already been through. As these continents undergo the world's next Great Transformation, it is a question, not just for them, but for all of us.

As pioneers of industrialism, it is arguably the West's collective duty to explore alternative models of social maturity. If we can decouple consumerism from the idea of a good life, we will have made the best possible use of 250 years' industrial experience. If, in addition, we address our demographic top-heaviness, not with robots, but by welcoming migrants, our shrinking populations will become a boon. Our greatest task now is to embrace our social maturity and to treat it not as a crisis but as an opportunity to rethink how we relate to time.

Beating the Clock

Thus, though we cannot make our sun / Stand still, yet we will make him run.
Andrew Marvell[22]

What is our relationship with time? In the West, the overriding feeling is that time is something of which we never have enough: we dash from one task to the next, gobbling down lunch while staring at our screens. Even those of us who could afford to take it easy complain of being 'money rich, time poor'.

Before the days of clocks and artificial light, things were very different. People had no choice but to live in synch with nature's rhythms. We live on a rotating planet, after all, so life among traditional peoples was arranged according to the ebb and flow of winds, tides, seasons and the alternations of day and night. Indigenous cultures generally had two essential measures of time: the cosmic rhythm of seasons and days and those associated with specific events such as harvesting and milling or domestic tasks such as making bread. Activities often stood for time itself. In Madagascar, for example, 'rice cooking' meant half an hour, 'maize roasting' fifteen minutes and 'the frying of a locust' a brief moment similar to our 'twinkling of an eye'. The passage of a few minutes in medieval England was reckoned according to how long it took to boil an egg.[23]

The fact that so many timings were linked to food is no accident, since most people spent much of their time engaged with it. As Tim Ingold notes, time in traditional societies was both task-oriented and social; far from being some abstraction against which everyday activities were measured, it embodied them. For many traditional people, indeed, time was so peripheral to life that they had no abstract notion of it at all, as was the case with the South Sudanese Nuer, as the anthropologist E. E. Evans-Pritchard explained:

> The Nuer have no expression equivalent to 'time' in our language, and they cannot, therefore, speak of time as though it were something which passes, can be wasted, saved and so forth. I do not think that they ever experience the same feeling of fighting against time or of having to coordinate activities with an abstract passage of time, their points of reference are mainly the activities themselves, which are of a leisurely character . . . Nuer are fortunate.[24]

As Evans-Pritchard suggests, one can't help but feel a certain longing for the Nuer's serenity with respect to time: a way of life further removed from our own hyper-speed, round-the-clock, sleep-deprived culture is hard to imagine. In the West such temporal freedom was of course shattered by the Industrial Revolution, with its division of life into work and leisure and Franklin's fateful idea that time is money. As the historian Lewis Mumford noted in *The Myth of the Machine*, it was the clock, not the train, that truly heralded the machine age. By allowing time and space to be quantified, Mumford argued, this 'paragon of automatons' became 'an integral part of the system of control that Western man spread over the planet'.[25] As we saw earlier, E. F. Schumacher also noted the way in which the commodification of time split us into producer and consumer – two incomplete halves of a creature enslaved by temporal logic.[26]

Modern gig workers know only too well where this division has led. Today, our economy demands not just that we measure time, but that we outrun it. High-frequency trading, just-in-time deliveries, algorithms and AI are just some of the outcomes of our digital quest for speed. Minimising time, like cutting labour costs, is a logical goal of capitalism. Like their Victorian counterparts, contemporary gig workers are not only paid slave wages, they are also monitored and penalised if their work rate slows.

The desire to outpace time isn't just making us miserable, it's also making us sick. Our circadian rhythms – biological systems that align our bodily cycles with earthly ones – are getting seriously upset by our twenty-four-hour lifestyles. Run by a central 'pacemaker' in the hypothalamus and set by light-sensitive cells in our eyes, our body clocks are designed to keep us aligned with the diurnal cycle, so that we are ready to get up in the morning, eat, digest, defecate, exercise and go to bed at night.[27] Every cell in us oscillates to this rhythm, as do all plant cells: if you live on a rotating planet, knowing when to sleep and wake is clearly handy. Yet thanks to our relentless existence, many of us live in a state of permanent 'social jet lag', as our bodies struggle to adjust to our artificial timekeeping. Lack of sleep is exposing us to the same health risks as our dodgy diet, from depression and obesity to diabetes, cancer, heart attacks and dementia.[28] Irregular hours are particularly bad, since we're always eating at the wrong time for our bodies. For this reason shift workers tend to be fatter than regular workers, and lack of sleep is making many of us fatter too, as the 'hungry hormone' ghrelin kicks in, turning us into doughnut-munching Homer Simpsons.[29]

Time's Arrow

> One cannot step twice into the same river.
>
> Heraclitus[30]

It is ironic that our frantic lives are partly driven by an economy based on the principle of infinite postponement. By offering a future reward in return for current sacrifice, capitalism places us on an invisible treadmill, living in constant hope of better times ahead. Like children yearning for Christmas or cartoon characters running against a static backdrop, we live in a state of permanent suspension, unable to enjoy what we've already got. For all its achievements, capitalism ultimately denies us happiness by preventing us from dwelling in time.

Set against this yearning for the future is an equally powerful desire to slow the flow of time, fuelled by the sense that its natural tendency is towards decline. The Garden of Eden is just one of many myths portraying a former state of bliss from which mankind has fallen, a trope cemented in our minds by the idea of a past golden age, first described by Hesiod in his *Works and Days*:

> The race of men that the immortals who dwell on Olympus made first of all were of gold. They lived like gods, with carefree heart, remote from toil and misery. Wretched old age did not affect them either, but with hands and feet ever unchanged they enjoyed themselves in feasting; beyond all ills, they died as if overcome by sleep.[31]

After this promising start, it all goes rapidly downhill: the golden age is followed by a 'much inferior' silver one in which witless men stay with their mothers to the age of a hundred before rapidly ageing and dying. Next comes a bronze age peopled by 'terrible and fierce' warriors who would have given Hobbes a run for his money, then, after a brief heroic interlude, comes the iron age of men, born to 'toil and misery by day and night'.[32]

This glass-half-empty view of our cosmic trajectory was adopted by all three Abrahamic religions, with the caveat that salvation may await somewhere down the line. Time was established as having started at a fixed point in the past (creation), moving through the present to some future apocalypse or day of judgement, at which point it would cease. Created by God (who was thus eternity), it was seen as a linear flow, sometimes called time's arrow, within which everyday experience was deemed to sit. In the nineteenth century, the second law of thermo-dynamics lent scientific heft to this apocalyptic view, stating that the progressive conversion of energy into ever lower forms (as when a burning log turns to ash) meant that time's flow was irreversible. This law of increasing entropy, the theory predicts, will eventually lead to the end of the universe, when all potential energy will have been used up and matter reaches a state of motionless equilibrium or 'heat death'.[33]

It's not hard to see how such theories feed into our sense of temporal unease. The idea that time has a beginning, middle and end maps easily on to our earthly lifespan from birth to death, as does the idea that its passage is not just irreversible, but destructive. One need only look in the mirror each morning to chart the inexorable decay, just as one can stare at an hourglass and watch time physically ebbing away.

So embedded is our linear perception of time, that it can be a shock to realise that not everyone thinks this way, yet, as with nature, our Western view of time is far from universally shared. Indian cosmic time, for example, is not seen as linear, but cyclical: an eternal cycle of destruc-tion and recreation echoed in the human realm by *samsara*, the wheel of life and death. In order to escape this endless cycle, one must lead a succession of virtuous lives in order to perfect one's soul and reach a state of stable, non-reincarnated bliss: the Buddhist nirvana, known to Hindus and Jains as *moksha*.

A Good Death

> Life, it is thanks to death that you are precious in my eyes.
>
> Seneca[34]

This very different cosmic view of time in India is mirrored, as we have seen, by a very different approach to life and death to that in the West. Physical conditions play a large part too: when life is generally short (in

India life expectancy is just sixty-nine, although up from forty-two in 1960), one has little choice but to view death with a certain pragmatism.[35]

The same was true of nineteenth-century Europe: while the average Briton today can expect to live to eighty-two, in 1800 just half made it past forty. Many died in their infancy: one in three British and German children died before the age of five, while in the US, the figure was close to one half.[36] The odds in medieval times were even worse, with famine, war and disease keeping average lifespans hovering in the low to mid-thirties. Death could come swiftly at any age, regardless of status, either from natural causes or in the form of some nameless disease that could carry you off within hours, such as the dreaded 'sweating sickness' that swept through Tudor England and did for Thomas Cromwell's wife and daughters.[37] Instead of battling death to the bitter end – a lost cause in most cases anyway – it's hardly surprising that people focused instead on preparing for the afterlife, using popular guides like the 1415 *Ars moriendi* (*Art of Dying*), which took them and their families through the various steps necessary for salvation.

While belief in an afterlife has been the commonest human response to mortality (incidentally producing some of our greatest art) it has by no means been universal. The other great tradition of dealing with death has been secular, and this began, as did so much else, with Socrates. For many Greeks, the manner of Socrates' death – which he met, as we saw, with impressive calm – was exemplary and inspiring. Today, we should say he was remarkably stoical, but the word hadn't yet been invented, since it came from the Stoics (named after the Stoa Poikile where they met), who formed a century after Socrates' death, yet considered him their teacher.[38]

Stoics believed that nature was ordered according to a divine logic (*logos*) aimed at the highest possible good.[39] Since no human could alter its course, a virtuous life thus consisted of obeying nature's laws and accepting one's fate unflinchingly. Misfortunes, said the Stoics, were just tests sent by the gods to help us develop fortitude, which we could do by mastering our emotions. As one might expect from such a chin-up attitude, command over death was held by the Stoics to be a crowning virtue. There was nothing to fear from death, they argued: it was mere nothingness, and since one would not be there to experience it, one would not suffer.

The Stoics, you may have noticed, were positively Epicurean in their view of death; indeed, both philosophical schools regarded Socrates as their ultimate teacher on all matters mortal. Epicurus' four key principles were summarised by his follower Philodemus as a 'tetrapharmakos', or four-part remedy: 'Don't fear god, don't worry about death, what is good is easy to get and what is terrible is easy to endure.'[40] Pain was easy to bear, said Epicurus, since minor ailments would soon pass, while major ones would quickly kill us. If we found ourselves in great pain, we could use our mental powers to overcome it, focusing our minds on pleasant memories. Lying on his own deathbed, Epicurus practised what he preached, insisting that the agonies he suffered (from a kidney stone) were easily outweighed by the pleasure he took in recalling happier times.

Such stoicism – for there is no other word for it – held a strong appeal for the down-to-earth Romans, who made it their dominant philosophy. In their unpredictable culture, love of fate (*amor fati*) was a useful skill, not least for the prominent Stoic Seneca, whose position as mentor to the Emperor Nero gave him plenty of it to deal with. Adversity was necessary to a good life, said Seneca, since without it one would never discover one's true worth. 'To be always happy and pass through life without any mental distress is to lack knowledge of one half of nature,' he wrote.[41] Coping with adversity not only made us stronger, it prepared us for our greatest challenge, facing our own demise.

Preparing for death should be easy, said Seneca, since we start dying the day we are born. To accept this was the key to a happy life, since once one had faced one's own mortality, one could live each day without fear, as though it were one's last, with the same pleasurable intensity. It was not the length of one's life that mattered, but the ability to live it to the full. 'To have lived long enough,' Seneca wrote, 'depends neither upon our years nor upon our days, but upon our minds.'[42] Rather than resist the passage of time, we should learn to savour its effects: 'Let us cherish and love old age,' he said, 'for it is full of pleasure if one knows how to use it. Fruits are most welcome when almost over; youth is most charming at its close; the last drink delights the drinker.'[43] Given that our human lifespan is inevitably short, furthermore, trying to prolong it was pointless:

> Place before your mind's eye the vast spread of time's abyss, and consider the universe; and then contrast our so-called human life with infinity: you will then see how scant is that for which we pray, and which we seek to lengthen.[44]

In AD 65, Seneca also had to practise what he preached, when, suspecting his involvement in a plot to kill him, Nero ordered his old tutor to take his own life, leaving the manner of death up to him. Gathering his friends and family around him and reminding them not to be distressed, Seneca calmly proceeded to open his veins and then carry on discussing the meaning of life, even as his own ebbed slowly away.[45]

Borrowed Time

What is striking about Seneca's writings is how closely they resemble those of modern thinkers such as Atul Gawande who seek to help us deal with our own mortality. Stoicism, in essence, is what Gawande is calling for, not just in order to assist us in facing death, but to help us lead better lives.

Western society is arguably the least stoical in history; our comfortable, risk-averse consumerist culture is aimed, after all, at removing all pain, suffering or effort (even that needed to peel a potato) from our lives, and thus any need for forbearance. Yet, as we have seen, the attempt to edit out such exertions and negatives hasn't made us any happier. On the contrary, expectation of a pain-free, serene existence merely prevents us from taking much pleasure in the comforts we enjoy. When did you last sigh in gratitude when you turned on a tap or flushed the loo? We've forgotten the cushion of convenience upon which our lives rest, and the fact that pain and effort are necessary conjuncts to joy and fulfilment.[46]

Seneca's writings feel relevant today because death, like food, is a constant. Although we live in very different times, the question that ultimately frames our lives – how to live in the face of mortality – is the same. Stoics, above all, were realists, which is why they can help us remain grounded in our runaway world. At the heart of their thinking was a familiar idea: that a good life was one lived in balance with nature. For the Stoics, this meant not just attuning oneself to the physical realm, but accepting one's place in time.

As Atul Gawande points out, this is partly to do with realising that the length of time we spend on earth doesn't just affect *us*. Gawande's Indian grandfather, for example, lived to the remarkable age of 110, remaining in charge of his family farm and making daily inspections of his fields on horseback right to the end. His grandfather's life sounds idyllic, said Gawande, until one considers its effect on the rest of his family: 'imagine how my uncles felt,' he wrote, 'as their father turned a hundred and they entered old age themselves, still waiting to inherit land'.[47]

As good political animals, the questions we must ask – how to live equitably and in balance with nature – clearly have a temporal dimension. If we were all to start living to 125, for example, our material demands would increase by 50 per cent (unless we starved ourselves in the process). Similarly, when we debate how to share our planet's resources fairly, who or what are we trying to share them with? All the humans and non-humans who currently inhabit the planet or our distant descendants? Such questions are central to the concept of sustainable development, first defined in the 1987 United Nations Brundtland Commission report *Our Common Future* as 'development that meets the needs of the present without compromising the ability of future generations to meet their own needs'.[48]

As anyone who has struggled to leave the last chocolate in the box will know, we're not wired to put off pleasure in the hope of some in the future. For better or worse, we prefer instant gratification, which is what our consumer-based industry is increasingly geared up to provide. It's hardly surprising that our hunter-gatherer psyches have led us into peril in such an environment. Honed for survival in a very different place and time, our tribal instincts don't extend easily to people or creatures we'll never meet, let alone to nameless billions yet to be born. Which is why, if Seneca were alive today, he would probably be hailed as an ecological guru. His advice to enjoy life while you can without trying to extend its natural limits is a form of temporal ecology. Accepting our own transience is both the key to living well and to making space for others to do the same. As Proudhon remarked, we only need our seat in the theatre for as long as it takes to watch the play; we don't need to own the whole building.[49]

Nunc est Bibendum

More than any other activity, eating expresses our transience, since it is the primary means by which our bodies transform. While we may live for many years, the atoms that constitute us are constantly changing, so that by the time we are adults, few if any of those with which we were born remain. This idea may seem disconcerting at first, until you think about what it means to eat, when it becomes obvious. Food is the material out of which our future bodies will be made, while our excretions contain those of our past. We are, quite literally, changeable feasts: when we eat, we digest our future selves, just as the food we consume is merely borrowed from the earth, to which it will eventually return.[50]

Eating and drinking are inherently Stoic acts, not just because they represent transience, but because they give us pleasure in the here and now. Feasting obeys Stoicism's central command, '*Carpe diem*' (Seize the day). When dining and quaffing, as Bernard Crettaz realised, we feel happy and alive – so much so that we might even contemplate death. Epicurean drinking vessels were often decorated with skulls for this reason, reminding revellers to enjoy life while they could. The motif travelled to Rome, where '*Nunc est bibendum*' (Now is the time to drink) was a favourite toast. Taken from a poem by Horace celebrating the demise of Cleopatra, the toast was drunk both in triumph at the death of an enemy and in inebriated defiance of one's own.

Defiance of death was key to the Roman feast of Saturnalia, during which Saturn, the god of agriculture and plenty, was said to return, briefly bringing back his golden age with him. Apart from the inevitable swigging and troughing, Saturnalia was a time of general misrule, during which slaves might insult their masters, wear their clothes and even be served by them at table. Held at the winter solstice, just as darkness was about to turn back to light, the feast represented perhaps the greatest defence against the dread of death: ritual laughter.[51] By over-turning hierarchies and celebrating bodily pleasures, Saturnalia (many of whose rituals morphed seamlessly into Christmas) was a necessary suspension of time during which human and cosmic rhythms became fused. As the world hovered between life and death, people made merry, merging the tragic and the comic into a seamless whole.

In medieval Europe, the saturnalian spirit lived on in the form of Carnival, a feast that retained enough ribaldry to betray its pagan roots. Held to coincide with various Church festivals throughout the year – most notably before the fasting of Lent – Carnival was a celebration of all things carnal, featuring the dual themes of meat and sex.[52] Alongside the roasting of joints and the feasting on pies, it included such rituals as cross-dressing and the mock ploughing of virgins in town squares, all accompanied by the predictable barrage of sausage-themed puns.

As Mikhail Bakhtin argued in *Rabelais and His World*, far from being trivial, such topsy-turvy bawdiness was vital to the meaning of Carnival, in which 'the serious and the comic aspects of the world and of the deity were equally sacred, equally "official"'.[53] Held at key moments in the seasonal calendar such as midwinter and the spring equinox, Carnival brought cosmic order, quite literally, down to earth. Its festivals were thus essentially about transformation – from dark to light, past to future, living to dead and back again. 'Carnival,' wrote Bakhtin, 'was the true feast of time, the feast of change, becoming and renewal.'[54]

Fleshly Depredations

As the theme of Carnival suggests, the tragi-comedy of being human lies in knowing that, while we eat and make love today, we will one day be food for worms. Being brought down to earth means precisely that: accepting one's place in the natural cycle of life, death and rebirth. What ritual laughter does, said Bakhtin, is break down our image of ourselves as immaculate and immutable, so that, like weathered rock, we can dissolve into the flow of life. 'To degrade,' he wrote, 'is to bury, to sow, and to kill simultaneously, in order to bring forth something more and better.'[55]

The medieval image of the grotesque body – as found in cathedral gargoyles and in the works of Bruegel and Bosch, or as described by Rabelais in his farting, belching heroes – embodies this idea. Such images don't express outward ugliness, said Bakhtin, but rather an inner truth. By celebrating degradation, they help us find our material place in time:

> The unfinished and open body (dying, bringing forth and being born) is not separated from the world by clearly defined boundaries; it is blended with the world, with animals, with objects. It is cosmic, it represents the entire material bodily world in all its elements.[56]

During the Renaissance, such themes took on a rather more moralistic guise in the form of *memento mori* (remember you must die), symbolic reminders of mortality such as a skull perched on one's desk or Caravaggio's provocatively rotten basket of fruit. One of the most celebrated manifestations of this tradition is the ghostly smear that floats across Holbein's 1533 double portrait *The Ambassadors*. Portrayed in anamorphic projection so that it can only be viewed from the side, the apparition reveals itself to be a human skull. The artistic conceit can seem puzzling or even macabre to us, yet to Holbein's contemporaries its message would have been clear: no matter your wealth and status, don't get above yourself; death will render us all equal.

The Dutch still-life paintings known as *vanitases* convey a similar message. Startlingly lifelike depictions of recently abandoned feasts, with crumbled bread, spilled goblets of wine and ruddy lobsters staring out in beady accusation, such images are products of the Dutch Golden Age and the guilt of its merchants as they struggled to reconcile their Puritan values with their embarrassing cornucopia of riches. The paintings capture both the essence of life and its transitory nature as few works do. There, just before us, is a half-peeled lemon dripping with juice, or a sparkling glass of white wine catching the light. We feel we could reach out and drink the liquid, and so join the illusion for which we can't help but yearn: that we could prolong pleasure for ever in suspended time.

Anthropocene

'Anthropocene' is the first fully antianthropocentric concept.
Timothy Morton[57]

As the twenty-first century gathers pace, our need to slow down gets ever clearer. Living slowly, as the Stoics realised, doesn't mean lack of excitement, it simply means living in the moment. This is, in essence, what the Slow Movement is all about: like Stoicism, it recognises that life is best lived, not in the past or future but in the present. Seneca even went as far as to say that humans are more fortunate than gods, since we know our days to be numbered and can thus experience life in its full here-and-now glory in a way that the immortals never will.

That 'now' is of course the Anthropocene, a man-made catastrophe of such awesome proportions that most of us shrink from even thinking about it. Yet admitting our complicity, Timothy Morton argues in *Dark Ecology*, is the first step towards a cure. We struggle to face our predicament, he says, because we recoil from the uncanniness of watching our past actions coming back to haunt us. We suffer from the 'nausea of co-existence' that comes with the realisation that, on a finite planet, nothing ever quite 'goes away'.[58]

Morton's solution is effectively a hybrid of Stoicism and Carnival: he suggests that we accept the absurdity of our condition and learn to laugh at it. One way of doing this, he suggests, is to recognise that our human era is a mere 'scintilla of geological time' that in cosmic terms is insignificant. 'From the point of view of the entropy at the end of the universe,' he asks, 'who cares about the Anthropocene?'[59] Climate change is a catastrophe, says Morton, but it is only the latest of a series of 'nested catastrophes' stretching back in time.[60] It comes after the Ice Age, which succeeded the asteroid that killed the dinosaurs, which followed the Great Oxygenation Event, the formation of the moon, and, back at the start of time itself, the Big Bang. All these events are still playing out: our planet moves in the Big Bang's aftermath, just as the air we breathe is made by the ongoing Great Oxygenation Event. Time is a series of nested events that form one long, ongoing present. It isn't linear, therefore; it's concentric.

Morton's suggestion echoes Seneca's advice that we imagine ourselves in the 'vast space of time's abyss'. Playing with our sense of time can help us accept our own mortality and the eventual demise of our species: the ultimate stage, one might say, of ecological thought. It can also help us to accept our messy, damaged present and to face up to the challenge of tackling climate change. As the recent upsurge of political action demonstrates, from the school climate strikers to Extinction Rebellion, many are ready to overcome their fear of cosmic time and seize the day.

The Order of Time

Facing up to time is hard, but it is also empowering: the necessary adjunct to growing up. Only by living fully in the present – by reconciling our own lifespan with cosmic time – can we act without fear. One of the reasons why contact with nature soothes us is that, when we go into a forest or gaze at mountains, we connect with a different order of time – we commune with beings that were here long before us and will be here long after we've gone. We lose ourselves in time and accept the fleetingness of our existence. We finally fall off our anthropocentric perch – which is, as any slapstick comedian will tell you, inherently funny.

Thinking oneself into non-existence may not be everyone's bag, but as Buddhists and Stoics have known for centuries, it is a highly effective way of reconciling oneself with time. Much of our Western malaise, as the ecologist Joanna Macey has suggested, stems from the way our concept of time exists independently of our material experience – a division recognised by Buddhists as a major source of *dukkha*, unhappiness.[61] For the thirteenth-century Japanese Zen master Dogen, such dualism could never exist, since objects existed only in time, and time was manifest only through objects; an indivisibility he expressed as *uji*, or 'time-being'. Time was a flow for Dogen, but not of the sort that passed from one moment to the next, but rather one that embodied existence itself.

A remarkable aspect of Dogen's metaphysical thought is the resemblance it bears to modern theoretical physics, in particular the field of quantum gravity. As the Italian theoretical physicist Carlo Rovelli explains in *The Order of Time*, what we think of as time's arrow is little more than a useful fiction. In reality, as Einstein realised, time is relative, moving at different speeds depending on where one is. Instead of being a neutral abstraction, therefore, time is the essence of our being. We *belong* in time, just as birds, stones, trees and mountains do, not because we share the same present, but because time is the force field through which we interact. As a result, says Rovelli, we would do better to think of the world, not as a collection of objects, but as a network of events.[62] To realise the 'ubiquity of impermanence', he says, is to discover the truth of existence. 'We understand the world in its becoming, not in its being.'[63]

Flow

Festina lente (Hasten slowly).
Roman proverb

Whatever our view of cosmic time, one thing we all know is that the terrestrial sort is not experienced as a constant. One need only compare, say, the feeling of standing in the wrong queue at the post office to that of staring into a lover's eyes to know this. Under certain circumstances – performing music, climbing a mountain or building a ship in a bottle – time can seem to stand still. This is what, as we have seen, Mihaly Csikszentmihalyi calls flow – the secular counterpart to meditation.[64]

Time stands still when we're engaged. Rather than measure out our days against a ticking clock, therefore, we're better off pottering in the garden, painting a picture or baking a cake. 'But,' you might object, 'what about all the stuff we've *got* to do? What about *work*?' That, of course, is precisely what Schumacher was trying to address when he railed against the arbitrary temporal divisions of capitalism. When work is meaningful – when it is task-oriented and social – it helps us transcend time, because it engages us. For this reason, time spent at home, where non-economic work such as cooking or gardening takes place, can be a haven of creative engagement.[65] Because it lies outside the cash economy, domestic life transcends commodified time.

One might argue – as a games-obsessed teenager might – that time can also stand still when one is waging fantasy war on a computer. This is perfectly true, but there is an essential difference between gaming and activities such as gardening or meditating that depend on no artificial stimulus. Although both kinds of activity can produce the experience of flow, the former does so by drawing us into a virtual world, while the latter achieve it by embedding us in the real one. Like those who enjoy a tipple, gamers get their highs through synthetic stimulation, while growers and meditators create their own. Gamers, one might say, consume their timelessness, while gardeners and monks produce theirs.

This distinction is key to why the expansion of leisure likely to result from the robotisation of work could be so problematic. As creatures of late capitalism, few of us have the skills to deal with endless leisure – after work, most of us spend our time consuming the rewards of our labour in the form of ready-cooked meals, shopping and entertainment. Yet no amount of consumption can make up for a meaningless life. In order to flourish, we need to feel useful, which means that we need to perform helpful tasks and create things. A life of pure *otium* can never make us happy; we also need *negotium* – we need to produce as well as consume – to be active rather than passive.

In his dystopian 1932 novel *Brave New World*, Aldous Huxley explored what life might be like should that logic be overturned. Set in a future World State in which all family ties are abolished and humans are bred in test tubes to fulfil their assigned place in the social hierarchy (from intellectual Alpha-pluses to Epsilon drudges), the book portrays a society in which intimacy and emotion are forbidden, and life is spent in an endless round of mindless consumption, games, entertainment and sex. Should anyone feel any incipient distress, the Controller Mustapha Mond explains, they can simply take a free dose of the government-approved drug *soma*:

> the old men have no time, no leisure from pleasure, not a moment to
> sit down and think – or if ever by some unlucky chance such a crev-
> ice of time should yawn in the solid substance of their distractions,
> there is always *soma*, delicious *soma*, half a gramme for a half-holiday,

a gramme for a weekend, two grammes for a trip to the gorgeous East, three for a dark eternity on the moon . . .[66]

Brave New World is arguably more relevant now than at any time since it was written. Today, as human genetic modification becomes possible, citizens and consumers are indistinguishable, robots threaten livelihoods and millions depend on antidepressants and opioids, our world and Huxley's feel uncomfortably close.[67]

Huxley's message is that happiness can't be created through the eradication of suffering, since they are two halves of one and the same thing. Nobody in *Brave New World* fears death, since life for its *soma*-dosed citizens has no meaning; deprived of all human emotion, they have little to lose. Happiness depends on its polar opposite, which is why Huxley famously claimed the right to be unhappy. Without the dark, he realised, there could be no light.

Autumn

> The knowledge of impermanence that haunts our days is their very fragrance.
>
> Rainer Maria Rilke[68]

Like most children, I used to love the summer but hated autumn. Keats's 'season of mists and mellow fruitfulness' held no joys for me. That sinking feeling, after a summer spent happily splashing on the beach or playing in the garden, when the dreaded 'back to school' displays started to appear in shops, with their grey uniforms and piles of notebooks and pens, felt like Sunday night multiplied a thousand times. Daylight fading, nights drawing in, the smell of dank, decaying leaves – all felt like harbingers of doom.

My dread of autumn, I now realise, wasn't just down to my dragon-like teachers; it echoed a deeper primal fear, the very same that once prompted ancient farmers to make sacrifices to the gods, Romans to drink to excess and our ancestors to build Stonehenge. Our lives revolve around the sun, and with each passing year, one has the sense of another chapter closing and an unwritten one about to begin.

These days, no doubt partly thanks to having left school, I take equal delight in every season. I can at last agree with Emerson that 'Each moment of the year has its own beauty.'[69] I love observing how the trees in the garden square at the end of my street change through the year: how pale and soft their leaves are when they first appear in spring, how bright emerald and full they are at the height of summer as I walk gratefully in their shade, how they rustle in autumn as they turn to tones of rust and yellow, and how their dark, naked forms in winter stand out against a pale sky. My favourite thing of all, however, is to drink in the crazily profuse, optimistic fluffiness of pink cherry blossom set off by the deep turquoise of the heavens. I can empathise with what the TV dramatist Dennis Potter said in an interview with Melvyn Bragg in 1994, just weeks before he died of cancer, describing how he felt when he looked out at his plum tree:

> It looks like apple blossom but it's white, and looking at it, instead of saying 'Oh that's nice blossom,' you know – last week, looking at it through the window when I'm writing, it is the whitest, frothiest, *blossomest* blossom that there ever could be; and I can see it, and things are both more trivial than they ever were and more important than they ever were, and the difference between the trivial and the important doesn't seem to matter, but the *nowness* of everything is absolutely wondrous, and if people could see that, you know, there's no way of telling you – you have to experience it – but the glory of it, if you like, the *comfort* of it, the reassurance – not that I'm interested in reassuring people, bugger that; the fact is that if you *see* the present tense, *boy* do you see it, and *boy* can you celebrate it.

Potter was moved because he knew the plum tree would outlive him: when it bloomed again the next year he would not be there to see it. Despite this, he could rejoice in the intensity of his feeling and, being a celebrated dramatist, was able to express how he felt.

Few of us experience many such intense moments, let alone manage to express them – most of our lives are spent in mundane routine – yet it is precisely by finding delight in such things that we can make our lives sing. And there is no better way of doing that than by reconnecting with food. Since I have become a micro-farmer on my London roof, my entire relationship with time has changed. Instead of being measured against work commitments and holidays, it is about deciding what plants to grow, buying seeds and compost, planting and nurturing, watering and supporting, harvesting, pickling, eating and sharing my beloved vegetables. For the first time in my life, my actions are directly bound to earthly rhythms, a bond that is as demanding as it is satisfying.

As Epicurus, Seneca and Buddha all realised, being present is the essence of a good life. It is also key to living happily within our means. In our hectic world, it is how we can become good political animals. As Aristotle, Rousseau and Kropotkin all understood, to live well in society we must be able to stand alone. Engaging, observing, noticing and responding are all qualities necessary both to politics and to gardening. They are about participation and involve a kind of letting go. Such skills are the essence of living in time: the key to both enjoying a good life and to having a long one.

In 1994, the US explorer Dan Buettner set out to investigate what he calls the world's Blue Zones: communities where people enjoy exceptional longevity.[70] The five self-contained groups he studied – in Sardinia, the Greek island of Icaria, Costa Rica, Okinawa and California – all contained not just some of the oldest people in the world, but also the healthiest and happiest. Determined to understand why this was, Buettner compared what the communities had in common, coming up with nine key traits.[71] These included natural environments that required people to exercise, whether by walking up hills, working in gardens or performing household tasks, as well as having a strong sense of community, family, belief and belonging. All the groups shared a strong sense of purpose (*ikigai*, in Japanese) and routines that helped to reduced stress, such as going for walks or taking a rest. Last but not least, they all enjoyed moderate, plant-based diets with perhaps (in the case of the Sardinians) a daily glass or two of wine.[72]

Apart from the fact that none of the Blue Zones lies within reach of an Amazon fulfilment centre, what is striking is that those who live

there seem remarkably content. By leading simple, meaningful, physical lives, they achieve what the full might of capitalism so often fails
to deliver. While nobody would advocate a return to village-based
agrarianism – which is what all the Blue Zones represent – there is
clearly much we can still learn from such communities. Much of it
comes down to the story of an encounter between some wealthy
Americans and an old Greek man sitting under a tree in an olive
grove, sipping ouzo and staring out to sea.[73] The Americans enquire
from the Greek whether he owns the olive grove, and when he says
that he does, they ask him why he doesn't hire people to harvest his
olives and export the oil. 'And why would I do that?' asks the man, to
which the Americans reply that he could get rich and do anything
that he liked. 'You mean,' says the man, 'that I could sit under this tree
and sip ouzo at sunset?'

Food and Time

> He who sees the present has seen all things.
> Marcus Aurelius[74]

Happiness is ephemeral; like time, it is not a commodity. If we are to
find happiness, it follows that we must reset our temporal horizons: find
some way of reconciling human and cosmic time. And I believe that
one thing that can help us to do this is staring up at us from our plates.

The substance that connects us to one another and to our world,
food is our ultimate timekeeper. The product of living, breathing creatures that evolved to the rhythms of tides and seasons, it is the daily dose
of biochemical energy that fills our bodies with life. It is also the focus
of the single ritual that every one of us still performs. It is this last aspect
of food – its ritual power – that is key to all the rest, since, if we really
want to learn to live in time, ritual is what we need.

Ritual, as Mircea Eliade explained in *The Sacred and the Profane*, is
human life experienced between two orders of time. The momentary
performance of saying Mass, founding a building, blessing a baby,
carving a joint, saying grace or singing 'Auld Lang Syne' is an act that

carries all other previous such actions within itself: it is the living embodiment of nested time. Before technology allowed humans to travel beyond the speed of sound or to ping dinner invites off satellites, ritual was the means by which our ancestors transcended time. By combining the secular and sacred temporal orders – for example by making a sacrifice before sharing a meal – rituals were the means by which people situated themselves in time. When we repeat such acts from our distant past, we unify the present with a vaster, cosmic order. Such unity, Eliade argued, is a way of arresting time itself, since 'by its very nature sacred time is reversible in the sense that, properly speaking, it is a primordial mythical time made present'.[75]

Ever since culture has existed, food has been central to such rituals. Nothing else combines the cosmic and domestic aspects of our life as powerfully as food. As the rhythm of our daily lives and the embodiment of life and death, food epitomises the dual meaning of 'mundane': a word that we use to mean 'boring, humdrum, everyday', yet whose deeper root is 'worldly, cosmic, of the universe' (from the Latin *mundus*). Few words express our cosmic disconnectedness so powerfully. The clue lies in the fact that our planet earth and the living earth that sustains us bear the same name. It reminds us that the edible stuff that we chew or pour unthinkingly down our gullets, the living things that we don't value and can't be bothered to think about, the substance that shapes our world and without which we wouldn't exist – this stuff called food can reconcile us with time.

Food's unique potential to heal us stems from the profound relationship that it represents between us and our world. The fact that we are blind to its power is due to our having forgotten what it really is. By learning to see food – and to see the world through it – we can find our true place in the natural order again. By valuing what we eat and knowing what it is, we can reconnect ourselves to each other and to our world. This is the true meaning of sitopia: using food to understand what it means to be human and how to coexist with our fellow humans and non-humans through time. By eating together consciously, we can be both grounded and transported, connected to a greater order of things. We can exist, for a period, both inside and outside time, and we can feel profoundly at home.

It's nearly midnight on 27 December as I write, two days after one of the world's most celebrated feasts. As someone who likes to party with the best of them, I am feeling right royally stuffed. Yet, as I sit here groaning, I love the idea that two billion others across the globe are feeling just like me. London over Christmas is blissful: many Londoners leave to celebrate with their families in the rest of the country – here as elsewhere, Christmas is a time to go back to your roots. The streets are quiet, news is sparse, my inbox is deliciously empty; the only thing missing on my festive wish list is a dusting of snow. Time, to the extent that it ever can in our frenetic world, really does stand still. Saturnalia lives on: when the north of the planet passes through its longest night, we all need this moment of celebration, of common, cosmic connection.

There are no easy answers to our human dilemma, yet no matter what obstacles lie ahead, food can be our guide. None of us existed before food: it preceded us, anticipates us, sustains us and will outlive us. The relationship that binds us to those we love and to our living world is, in the end, our greatest hope.

Acknowledgements

If it takes a village to raise a child, it takes a community to write a book. *Sitopia* would not exist were it not for the help, support and inspiration of too many people to mention here, so while restricting myself to thanking those most directly involved, I should like to acknowledge all those whose lively engagement with my subject over the years has played such a big part in shaping this book.

First of all, my heartfelt thanks go to my wonderful editor Poppy Hampson, without whom this book would most certainly not exist, since it was she who first took me on as an unproven author for my first book, *Hungry City*, and who agreed to publish this one on the basis of a scrappy drawing and much arm-waving in a hot Italian restaurant back in 2011. Poppy's infinite patience, calm wisdom and impeccable judgement have all been invaluable in helping me write this book, and her steadfast belief in it has kept me going through the often arduous process of bringing it to life. I count myself among the luckiest of writers to work with her. My warmest thanks also go to the rest of the team at Chatto & Windus, in particular to Greg Clowes for his cheery enthusiasm and always prompt and pertinent advice, and to Kris Potter for his beautiful cover design. I should also like to thank my fantastic agent Jonny Pegg, who took me on as a rookie writer when my ideas were still vague at best, and stuck by me through the gruelling process of finding an editor with similarly pliant vision. Jonny's unfailing enthusiasm for my work has been a great fillip over the fifteen years we have worked together.

My deepest thanks go to the wonderful Miriam Escofet, who so generously made time in her hectic schedule to draw the exquisite frontispiece for this book. Her drawing is loosely based on the original sketch from which *Sitopia* came, so it is magical to see it transformed by her incomparable hand into such a fabulous vision. I should also like to thank Georgie Lowe for her impeccable 'dictionary-style' sitopian definition, which appears on the front cover.

The arguments in *Sitopia* have evolved over many years and through countless conversations. My warmest thanks go to all those who have not only engaged with the ideas but have so generously read the drafts to give me the benefit of their considerable wisdom: Charisse Amand, David Bass, Petra Derkzen, Karen Gilbert, Trine Hahnemann, Niels-Peter Hahnemann, Will Hancox, Robert Kennett, Alex Laird, Georgia Lowe, Rowan Moore, Richard Nightingale, Stanley Steel, Brian Vermeulen and Stephen Witherford. My special thanks go to David Bass, who, as with *Hungry City*, has been my most constant support and staunchest critic, in the nicest and most loving of ways. I should also like to give special thanks to Will Hancox, whose extraordinary mind and willingness to engage with the emergent themes of *Sitopia* have been instrumental in helping me develop its core ideas. Our decade-long conversation has not only been great fun, but has provided the chief compost for this book, many of whose arguments were first formed or tested with him across my kitchen table.

My grateful thanks also go to all those whose enthusiasm and encouragement have contributed so much to shaping the book: Cany Ash, Louise Ash, Steve Bass, Cressida Bell, Claire Bennie, Sarah Bilney, Peter Carl, Nat Chard, Lulu Chivers, Geoff Crook, Kath Dalmeny, Chris Dawe, Miriam Escofet, Harriet Friedman, Peter Gill, Claire Hartten, Thor Hartten, Nick Horsley, Salima Ikram, Jesus Jimenez, Tim Lang, Elisabeth Luard, Helen Mallinson, Michael Mallinson, Patricia Michelson, Kevin Morgan, Nick Morgan, Mohsen Mostafavi, Juliet Odgers, Eric Parry, Arthur Potts-Dawson, Claire Pritchard, Charlie Pye-Smith, Wendy Pullan, Cathy Runciman, Polly Russell, Joseph Rykwert, Robert Sakula, Dan Saladino, David Sawer, Geoff Tansey, Jeremy Till, Nick Warner, Sarah Wigglesworth, Han Wiskerke and Mike Yezzi. I should also like to give special thanks to Berte Fisher-Hansen and Peter Schultzer of Crossway Film for their beautiful film and to everyone at Kilburn Nightingale – including extended family – for their amazing support and numerous feisty debates in improbable places.

With a subject as broad as that of *Sitopia*, I have inevitably had to rely on others' expertise to fill in the gaps, so I am deeply grateful to all those who have so generously given their time to share their specialist knowledge with me: Richard Ballard, Martin Crawford, Steven Dring, Claude Fischler, Ben Flanner, Miles Irving, Alex Laird, Diana Lee-Smith, Charles Michel, Olivier Raevel, Charles Spence and Helena Tervo. I should also like to thank Thomas Bout and Antoinette Guhl for so kindly arranging my trip to Rungis and to Robin Harford for so generously sharing his remarkable wisdom on a memorable walk among the wild edibles of Devon.

Since straying into the world of food from my previous architectural habitat some twenty years ago, I have met numerous inspiring people whom I am now delighted to call my extended food family. Among these are a handful whose shared enthusiasm has not only blossomed into close friendship, but who have opened up entire new worlds for me: Claire Hartten and the wonderful 'Green Rabbits' of New York, Trine Hahnemann, who introduced me to her fantastic food network in Denmark, and Peter de Rooden and Arno van Roosmalen of Stroom den Haag, whose early interest in *Hungry City* led to its Dutch translation and a host of connections that have been formative in my subsequent work and continue to this day. Back in London, I remain grateful to Sarah Bilney and Patricia Michelson of La Fromagerie for providing the warmest, loveliest and most impeccably ethical food home anyone could wish for.

Last but not least, I should like to thank two special people who were alive when I began writing *Sitopia* but are sadly no longer with us: my inspirational teacher, Dalibor Vesely, who together with Peter Carl sketched out a world for me forty years ago that remains the bedrock of my thinking, and my wonderful father, Stanley Steel, whose deep wisdom, love and unparalleled sense of humour I draw on every day.

List of Illustrations

Frontispiece: *An Allegory of Sitopia* (2019) by Miriam Escofet

1. Albrecht Dürer, *Adam and Eve* (1504), The Met, New York (public domain)
2. Pieter Bruegel the Elder, *The Land of Cockaigne* (1567), Alte Pinakothek, Munich (© M455JG; Art Library, Mr John Harper / Alamy Stock Photo)
3. Anonymous, *Family eating their midday meal at home* (1946) (© Walter Sanders / The LIFE Picture Collection via Getty Images)
4. Anonymous, *Les Halles, Paris* (*c.* 1900) (© Culture Club / Getty Images)
5. *Très Riches Heures du Duc de Berry* (1411–1416), Musée Condé, Chantilly (© MMGYAD; The Picture Art Collection, Paul Fearn / Alamy Stock Photo)
6. *Excavated Apple Tree with Roots* (1970s), East Malling Research Station, Kent (courtesy of NIAB EMR)
7. Juan Sánchez Cotán, *Still Life with Game, Vegetables and Fruit* (1602), Prado, Madrid (© Photographic Archive Museo Nacional del Prado)

Notes

1 Food

1   https://www.pidgeondigital.com/talks/technology-is-the-answer-but-what-was-the-question-/

2   Winston Churchill, 'Fifty Years Hence', *Strand Magazine*, 1931: 'With a greater knowledge of what are called hormones, i.e. the chemical messengers in our blood, it will be possible to control growth. We shall escape the absurdity of growing a whole chicken in order to eat the breast or wing, by growing these parts separately under a suitable medium.' https://www.nationalchurchillmuseum.org/fifty-years-hence.html

3   In 2010, the UN admitted that a key finding in *Livestock's Long Shadow*, also quoted in *Cowspiracy*, that livestock production contributed 18 per cent of global greenhouse gas emissions, was inaccurate. The revised figure is 14.5 per cent. http://news.bbc.co.uk/1/hi/8583308.stm; http://www.fao.org/docrep/010/a0701e/a0701e00.HTM; https://www.thelancet.com/commissions/EAT

4   Due to the protein conversion efficiencies of various livestock. See Vaclav Smil, *Enriching the Earth*, MIT Press, 2004, p.165.

5   'Towards Happier Meals in a Globalised World', www.worldwatch.org, 3 February 2014.

6   Raj Patel, *The Value of Nothing: How to Reshape Market Society and Redefine Democracy*, London, Portobello, 2009, p.44.

7   Jonathan Safran Foer, *Eating Animals*, Penguin, 2009, pp.92–3.

8   https://www.sentienceinstitute.org/us-factory-farming-estimates

9   M. Bar-On Yinon, Rob Phillips and Ron Milo, 'The biomass distribution on Earth', *PNAS*, 19 June 2018, 115 (25) 6506–11, https://www.pnas.org/content/115/25/6506

10   https://www.theguardian.com/sustainable-business/fake-food-tech-revolutionise-protein

11   https://money.cnn.com/2018/02/01/technology/google-earnings/index.html

12   http://www.fao.org/docrep/018/i3107e/i3107e03.pdf

13   The numbers fluctuate: for the latest stats see: http://www.fao.org/hunger/en/; http://www.who.int/mediacentre/factsheets/fs311/en/

14   Marion Nestle, *Food Politics*, University of California Press, 2002, p.13.

15   Tristram Stuart, *Waste: Uncovering the Global Food Scandal*, London, Penguin, 2009, p.188.

16   Ibid., p.193.

17   *Livestock's Long Shadow*, UN Food and Agriculture Organisation, Rome 2006.

18   http://www.nytimes.com/2013/06/16/world/asia/chinas-great-uprooting-moving-250-million-into-cities.html?_r=1&, accessed 6 March 2014.

19   Malcolm Moore, 'China now eats twice as much meat as the United States', *Daily Telegraph*, 12 October 2012.

20   http://culturedbeef.net

21   Virginia Woolf, *A Room of One's Own* (1928), Bloomsbury, 1993, p.27.

22   As the US psychologist Daniel Kahneman has pointed out, such decisions are often made more unconsciously than we think. See Daniel Kahneman, *Thinking, Fast and Slow*, Penguin, 2011, pp.39–49.

23   Richard Layard, *Happiness: Lessons From a New Science*, Penguin, 2005, p.3.

24   Edith Hamilton and Huntingdon Cairns (eds), *Plato: The Collected Dialogues*, Princeton University Press, 1987, p.23.

25   Aristotle, *The Nicomachean Ethics*, J. A. K. Thomson (trans.), Penguin, 1978, p.63.

26   Op. cit., p.109.

27   Douglas Adams, *The Hitchhiker's Guide to the Galaxy*, Pan Books, 1979, pp.135–6.

28   For a detailed discussion of how this works, see Shoshana Zuboff, *The Age of Surveillance Capitalism*, Profile Books, 2019.

29   Jared Diamond, *Collapse: How Societies Choose to Fail or Survive*, Penguin, 2006, p.11.

30   Jean Anthelme Brillat-Savarin, *The Physiology of Taste* (1825), Penguin, 1970, p.54.

31   See the Introduction, p. 2.

32   http://www.dailymail.co.uk/femail/food/article-1341290/The-adored-mother-meals-I-hated-The-evil-stepmum-cooked-like-dream-And-food-shaped-bittersweet-childhood-TV-chef-Nigel-Slater.html

33   Brillat-Savarin, op. cit., p.13.

34   Charles Darwin, *On the Origin of Species* (1859), Oxford World's Classics, 2008, pp.50–1.

35   The chemical symbol for potassium is K, from the medieval Latin *kalium* – potash.

36   See Smil, op.cit., p.160.

37   Genesis 1: 29

38   Genesis 2: 17

39   See Reay Tannahill, *Food in History*, Penguin, 1973, pp.105–9.

40   Epicurus, *The Art of Happiness*, George K. Strodach (trans.), Penguin, 2012, p.183.

41   Ibid., p.61.

42   Abraham Maslow, *Towards a Psychology of Being* (1962), Wilder Publications, 2011, p.27.

43   Ibid., p.36.

44   A strategy that tends to backfire, Maslow noted, since few people appreciate being treated as 'need-gratifiers', rather than as themselves (ibid., p.37).

45. Ibid., p.33.

46   Mihaly Csikszentmihalyi, *Flow: The Psychology of Optimal Experience*, Harper Perennial, 2008.

2 Body

1   Brillat-Savarin, op. cit., p.162.

2   http://www.annualreports.co.uk/Company/weight-watchers-international-inc

3   https://www.globenewswire.com/news-release/2019/02/25/1741719/0/en/United-States-Weight-Loss-Diet-Control-Market-Report-2019 Value Growth-Rates-of-All-Major-Weight-Loss-Segments-Early-1980s-to-2018–2019-and-2023-Forecasts.html

4   https://www.huffingtonpost.co.uk/2016/03/10/majority-brits-are-on-a-diet-most-of-the-time_n_9426086.html

5   https://nypost.com/2018/09/26/nobody-eats-three-meals-a-day-anymore/

6   https://harris-interactive.co.uk/wp-content/uploads/sites/7/2015/09/HI_UK_FMCG_Grocer-report-bagged-snacks-February.pdf

7   https://news.stanford.edu/news/multi/features/food/eating.html

8   According to some studies, the mantle has now passed to Mexico, directly as a result of having adopted American-style food.

9   The UK's 'special relationship' with the US made us particularly susceptible to its fast-food culture.

10   Harold McGee, *On Food and Cooking, The Science and Lore of the Kitchen*, Charles Scribner, New York, 1984, p.561.

11   Jean-Jacques Rousseau, *Émile*, 1762: http://www.gutenberg.org/cache/epub/5427/pg5427.html

12   Charles Spence and Betina Piqueras-Fiszman, *The Perfect Meal: The multi-sensory science of food and dining*, Wiley Blackwell, 2014, p.201.

13   Marcel Proust, *Remembrance of Things Past*, Vol. 1. *Swann's Way*, Chatto and Windus, 1976, p.58.

14   Bee Wilson, *First Bite: How We Learn to Eat*, 4th Estate, 2016, p.117.

15   On a BBC2 *Horizon* programme: https://www.telegraph.co.uk/culture/tvandradio/9960559/Horizon-The-Truth-About-Taste-BBC-Two-review.html

16 http://www.sciencemag.org/content/343/6177/1370

17 Quoted in an interview with the author.

18 Ibid., p.116.

19 Harold McGee, op. cit., p.562.

20 Brillat-Savarin, op. cit., p.13.

21 Charles Darwin, *The Descent of Man* (1879), Penguin, 2004, p.68.

22 http://www.scientificamerican.com/article/thinking-hard-calories/

23 Richard Wrangham, *Catching Fire: How Cooking Made us Human*, Profile Books, 2009, pp.109–13.

24 Gaston Bachelard, *The Psycholanalysis of Fire*, Beacon Press, 1968, p.7.

25 The artist Rudolph Zallinger said he had no such intention.

26 *The Surgeon General's Vision for a Healthy and Fit Nation 2010*, U.S. Department of Health and Human Services.

27 Edward O. Wilson, *The Social Conquest of Earth*, New York, Liveright Publishing Corporation, 2012, p.7.

28 Isaiah, 40: 6: 'All flesh is grass, and all the goodliness thereof is as the flower of the field.'

29 Although the curative effects of citrus fruits were known, it was not until the British naval surgeon James Lind carried out a clinical trial in 1747 that their capacity to prevent scurvy was proven, leading to British ships carrying the fruit as standard issue and their crews being nicknamed Limeys.

30 Michael Pollan, *In Defence of Food*, Allen Lane, 2008, p.117.

31 Although increasing numbers of people around the world, including the Chinese, now drink milk, the majority are lactose intolerant.

32 Ibid., p.102.

33 Professor Ralph C. Martin, University of Guelph, speaking at the Toronto Food Policy Council, 20 October 2011.

34 See Tim Spector, *The Diet Myth: The Real Science Behind What We Eat*, Weidenfeld & Nicolson, 2015, pp.118–22.

35 Michael Pollan, *The Omnivore's Dilemma: The Search for a Perfect Meal in a Fast-Food World*, Bloomsbury, 2006, p.84.

36 *The Big Bang Theory*, CBS, Season 1, Episode 4.

37 Graham Harvey, *We Want Real Food: Why Our Food is Deficient in Minerals and Nutrients and What We Can Do About It*, Constable and Robinson, 2006, p.52.

38 For a detailed discussion of the benefits of pasture-fed beef, see ibid., pp.82–99.

39 Ibid., p.95.

40 Quoted on the *Food Programme*, Radio 4, 2 November 2015.

41 The definition of ultra-processed food was first made by a team led by Professor Carlos Monteiro at the University of São Paulo in Brazil and is known as the Nova classification.

42 https://www.theguardian.com/science/2018/feb/02/ultra-processed-products-now-half-of-all-uk-family-food-purchases

43 https://www.theguardian.com/science/2018/feb/14/ultra-processed-foods-may-be-linked-to-cancer-says-study

44   Carlo Petrini, *Slow Food: The Case for Taste*, Columbia University Press, 2001, p.10.

45   Defined as somewhere you have to walk more than 500 metres to find a source of fresh food.

46   *Jamie's School Dinners*, Channel 4, 2005.

47   George Orwell, *The Road to Wigan Pier* (1937), Penguin, 2001, p.92.

48   For a detailed discussion of the way in which the American food system works, see Marion Nestle, *Food Politics: How the food industry influences nutrition and health*, University of California Press, 2002, pp.1–18.

49   news.yale.edu/2013/11/04/fast-food-companies-still-target-kids-marketing-unhealthy-products/

50   https://www.cdc.gov/nchs/data/hus/2018/021.pdf

51   http://www.gallup.com/poll/163868/fast-food-major-part-diet.aspx

52   Brenda Davis, 'Defeating Diabetes: Lessons From the Marshall Islands', *Today's Dietitian*, Vol.10, No. 8, p.24

53   http://www.dailymail.co.uk/health/article-2301172/Fattest-countries-world-revealed-Extraordinary-graphic-charts-average-body-mass-index-men-women-country-surprising-results.html

54   Bové was protesting against US sanctions imposed on Roquefort cheese (which he made) in retaliation for the European Community refusing to allow US hormone-induced beef into Europe.

55   http://www.aboutmcdonalds.com

56   http://www.telegraph.co.uk/news/worldnews/europe/france/10862560/French-town-protests-to-demand-McDonalds-restaurant.html

57   Andy Warhol, *The Philosophy of Andy Warhol* (1975), Harvest, 1977.

58   Oscar Wilde, *Lady Windermere's Fan* (1892), Act I., Methuen & Co, 1917, p.21.

59   http://www.scientificamerican.com/article/gut-second-brain/

60   Paul J. Kenny, 'Is Obesity an Addiction?', *Scientific American*, 20 August 2013: http://www.scientificamerican.com/article/is-obesity-an-addiction/

61   Ibid.

62   Ibid.

63   *Horizon*, 'The Truth about Fat', BBC2, 21 March 2012.

64   http://www.poverty.org.uk/63/index.shtml

65   The French spend two hours thirteen minutes a day on average eating and drinking, more than any other nation and more than twice as much as Americans, who spend just one hour and one minute: https://www.thelocal.fr/20180313/french-spend-twice-as-long-eating-and-drinking-as-americans

66   Paul Rozin, Abigail K. Remick and Claude Fischler, 'Broad themes of difference between French and Americans in attitudes to food and other life domains: personal versus communal values, quantity versus quality, and comforts versus joys', *Frontiers in Psychology*, 26 July 2011.

67   Ibid., p.8.

68   Ibid., p.2.

69   Private correspondence with Claude Fischler.

70   A personal motto of Mies.

71   Ibid. p.18.

72   Ibid. p.73.

73   Harvey Levenstein, *Revolution at the Table: The Transformation of the American Diet*, University of California Press, 2003, p.93.

74   Harold McGee, op. cit.,p.283.

75   Quoted in ibid, p.246.

76   Kellogg believed the diet came directly from God, via a communication to the Church's leader Ellen White in 1863.

77   In fact, the name on your cereal packet refers to John's brother Will, who after a rift with his brother set up the Battle Creek Toasted Corn Flake Company, the firm we now know as Kellogg's.

78   Pollan, *In Defence of Food*, p.45.

79   Ibid, p.22. Pollan credits Australian sociologist Gyorgy Scrinis with the term.

80   http://nutribase.com/fwchartf.html

81   'Americana: The Theory of Weightlessness', *Time* magazine, 21 November 1960: http://content.time.com/time/magazine/article/0,9171,874185,00.html (the Schmoo was a cartoon character shaped like a bowling pin).

82   http://www.independent.co.uk/life-style/health-and-families/features/the-science-of-saturated-fat-a-big-fat-surprise-about-nutrition-9692121.html

83   Sugar itself was gradually replaced by the sweeter, cheaper and harder to digest high-fructose corn syrup (HFCS), first developed by Japanese scientists in 1971.

84   https://experiencelife.com/article/a-big-fat-mistake/

85   The name of Yudkin's book, published in 1972.

86   http://www.telegraph.co.uk/news/celebritynews/6602430/Kate-Moss-Nothing-tastes-as-good-as-skinny-feels.html

87   http://centennial.rucares.org/index.php?page=Weight_Loss

88   http://www.dailymail.co.uk/health/article-2117445/Women-tried-61-diets-age-45-constant-battle-stay-slim.html

89   http://sheu.org.uk/content/page/young-people-2014

90   https://www.theguardian.com/society/2019/feb/15/hospital-admissions-for-eating-disorders-surge-to-highest-in-eight-years

91   So obsessed do some dieters become with following such 'healthy' regimes (which often involve leaving out entire food groups, such as gluten or dairy) that they can not only become malnourished, but develop orthorexia nervosa, a condition similar to obsessive compulsive disorder.

92   https://www.npd.com/wps/portal/npd/us/news/press-releases/the-npd-group-reports-dieting-is-at-an-all-time-low-dieting-season-has-begun-but-its-not-what-it-used-to-be/

93   Rob Rhinehart, *How I Stopped Eating Food*, posted 13 February 2013 on his now-defunct blog, Mostly Harmless. (*Mostly Harmless* was the fifth book in Douglas Adams' series *Hitchhikers Guide to the Galaxy* and refers to the entry in the guide for planet earth).

94   Ibid.

95   http://www.economist.com/blogs/babbage/2013/05/nutrition

96   Rob Rhinehart, Mostly Harmless, 25 April 2013.

Notes

97 https://www.ft.com/content/77666780-4daf-11e6-8172-e39ecd3b86fc;
    https://www.newyorker.com/magazine/2014/05/12/the-end-of-food
98 R. Buckminster Fuller, *Nine Chains to the Moon* (1938), Anchor Books, 1973,
    pp.252–9.

3 Home

1   Gaston Bachelard, *The Poetics of Space*, Beacon, 1969, p.4.
2   Ibid., p.14.
3   Wrangham, op. cit., p.157.
4   In a famous experiment carried out during the 1960s and 70s at Stanford
    University, Psychology Professor Walter Mischel offered some four-year-old
    children a choice between an immediate sweet treat (such as a marshmallow
    or Oreo cookie) or a bigger treat (two cookies), should they be able to resist
    the temptation to eat for 15 minutes. As young adults, Mischel found, the
    children who had been able to resist the temptation had considerably better
    life outcomes than those who had not. See Kahneman, op. cit., p.47.
5   See Margaret Visser, *The Rituals of Dinner*, Penguin, 1991, p.91.
6   Brillat-Savarin, op. cit., p.55.
7   https://www.theguardian.com/society/2018/may/23/the-friend-effect-why-
    the-secret-of-health-and-happiness-is-surprisingly-simple
8   The oxytocin surge set off serotonin and dopamine too. See Paul J. Zak, *The
    Moral Molecule,* Corgi, 2012, pp.28–32 and pp.95–100.
9   Brillat Savarin, op. cit., p.163.
10 Some locals object to the 'filthy stench in the air on Sunday evenings'. See
    David Howes (ed.), *Empire of the Senses: The Sensual Cultural Reader*, Berg,
    2004, p.232.
11 Judith Flanders, *The Making of Home*, Atlantic Books, 2014, p.185.
12 See Joseph Rykwert, *The Idea of a Town*, Faber and Faber, 1976, p.168.
13 Ibid., pp.121–6.
14 Colin Turnbull, *The Forest People*, Simon and Schuster, 1962, p.14.
15 Ibid., p.92.
16 Ibid., p.26.
17 Quoted in Tim Ingold, *The Perception of the Environment, Essays on Livelihood,
    Dwelling and Skill*, London and New York, Routledge, 2011, p.21.
18 Ibid., p.22.
19 Ibid., p.23.
20 Jean-Jacques Rousseau, *The Social Contract* (1762), Penguin, 2004, p.2.
21 The relationship is often to be found in language, as in the Anglo-Saxon *heorp*,
    hearth, which also stood for the whole house. See Flanders, op. cit., p.56.
22 Wrangham, op. cit., pp.138–9.
23 Ibid., pp.135–6.
24 Wilson, op. cit., p.44.
25 Ibid., p.17.

26   Ibid.

27   See Yuval Noah Harari, *Sapiens: A Brief History of Humankind*, Harvill Secker, 2014, pp. 20–1.

28   Jean-Jacques Rousseau, *The Social Contract and The First and Second Discourses*, Susan Dunn (ed.), Yale University Press, 2002, p. 120.

29   Wilson, op. cit., p. 93.

30   In fact, some groups had already started to settle, especially those close to a good source of food such as a river. See Tom Standage, *An Edible History of Humanity*, Atlantic Books, 2010, pp. 20–1.

31   Ibid., pp. 13–15.

32   The !Kung bushmen of the Kalahari spend between twelve and nineteen hours a week collecting food, leaving plenty of time for other activities. See ibid., p. 16.

33   Ingold, op. cit., pp. 323–4.

34   Standage, op. cit., p. 18.

35   Ibid.

36   Ibid.

37   Jared Diamond, *Guns, Germs and Steel: A Short History of Everybody for the Last 13,000 years*, Vintage, 2005, p. 142.

38   Estimates by Philip M. Hauser, quoted in Norbert Schoenauer, *6000 Years of Housing*, W.W. Norton and Co., 1981, p. 96.

39   The anthropologist Nurit Bird-David has noted that while forest dwellers typically see the forest as a parent that gives its bounty unconditionally, farmers see the land as an entity that yields its bounty reciprocally, in return for favours rendered. See Ingold, op. cit., p. 43.

40   Hesiod, *Works and Days* (701–702), in *Hesiod, Theogony and Works and Days*, M. L. West (trans.), Oxford World Classics, OUP, 2008, p. 58.

41   Aristotle, *The Politics*, T.A. Sinclair (trans.), Penguin, 1981, p. 56.

42   Ibid., p. 59.

43   Ibid., (404–412), p. 49.

44   Ibid., 7.35, p. 453.

45   Ibid., 7.3, p. 441.

46   Ibid., 10.12, p. 479.

47   Ibid., p. 85.

48   Ibid., p. 4.

49   Flanders, op. cit., p. 28.

50   By wiping out around one third of the population of Europe, the Black Death was a major factor in the lessening of feudal power: with few workers to go round, lords were forced to treat their peasants better.

51   David J. Kerzer and Marzio Barbagli (eds.), *Family Life in Early Modern Times, 1500–1789, The History of the European Family*, Vol. 1, Yale University Press, 2001, pp. 39–40.

52   Flanders, op. cit., p. 34.

53   By contrast, wives in early marriage societies were generally forbidden to remarry.

54 Ibid., p.48.

55 Peter Laslett, *The World We Have Lost Further Explored*, Routledge, 1994, p.1.

56 Ibid., p.3.

57 Flanders, op. cit., p.33–4.

58 Laslett, op. cit., p.4.

59 William Blake, *Milton* (1804–8), quoted in Humphrey Jennings, *Pandaemonium 1660–1886, The coming of the machine as seen by contemporary observers* (1985), Icon Books, 2012, p.127.

60 W. G. Hoskins, *The Making of the English Landscape*, Pelican, 1955, p.185.

61 See Frank E. Huggett, *The Land Question*, Thames and Hudson, 1975, pp.21–4.

62 The war had kept prices artificially high.

63 George Crabbe, *The Village*: http://www.gutenberg.org/files/5203/5203-h/5203-h.htm

64 The seventeenth-century practice of 'putting out' – in which cotton merchants brought raw materials directly to farmhouses to be spun and woven, returning later to collect the finished cloth – gave rural households a further economic boost.

65 Friedrich Engels, *The Condition of the Working Class in England* (1845), Penguin, 2009, p.92.

66 Eric Hobsbawm, *The Age of Revolution 1789–1848*, Abacus, 1962, p.66.

67 Engels, op. cit., p.167.

68 John Ruskin, *Sesame and Lilies* (1865): http://www.gutenberg.org/cache/epub/1293/pg1293-images.html

69 John Burnett, *A Social History of Housing 1815–1970*, Newton Abbott, David and Charles, 1978, p.185.

70 Theodore Zeldin, *An Intimate History of Humanity*, Vintage, 1998, p.370.

71 Burnett, op. cit., pp.188–9.

72 Charles Pooter was the fictional bank clerk in George and Weedon Grossmith's 1892 comedy of suburban life, *Diary of a Nobody*.

73 Wrangham, op. cit., p.151.

74 Burnett, op. cit., p.145.

75 The total is uncertain, since many domestic workers not included in official figures shifted to other jobs during the war. See Gail Braybon, *Women Workers in the First World War*, Routledge, 1989, p.49.

76 Walter Long, president of the Local Government Board: ibid., p.215.

77 John Burnett, 'Time, place and content: the changing structure of meals in Britain in the 19th and 20th centuries', in Martin R. Schärer and Alexander Fenton (eds), *Food and Material Culture*, East Linton Scotland, Tuckwell Press, 1998, Ch. 9, p.121.

78 'Hitler threatens Europe,' one advert ran in *The American Home*, 'but Betty Haven's boss is coming to dinner and that's what *really* counts.' See Harvey Levenstein, *Paradox of Plenty: A Social History of Eating in Modern America*, Oxford University Press, 1993, p.32.

79 Ibid., p.143.
80 For a detailed discussion of the development of kitchen design, see Carolyn Steel, *Hungry City: How Food Shapes Our Lives*, Chatto & Windus, 2008, pp.155–200.
81 http://www.striking-women.org/module/women-and-work/post-world-war-ii-1946–1970
82 https://www.census.gov/newsroom/press-releases/2016/cb16-192.html; https://www.ons.gov.uk/peoplepopulationandcommunity/birthsdeathsand marriages/families/bulletins/familiesandhouseholds/2017
83 https://www.bls.gov/opub/ted/2017/employment-in-families-with-children-in-2016.htm; https://www.ons.gov.uk/employmentandlabour-market/peopleinwork/employmentandemployeetypes/articles/familiesand thelabourmarketengland/2017
84 https://www.theatlantic.com/magazine/archive/2010/07/the-end-of-men/308135/
85 https://qz.com/1367506/pew-research-teens-worried-they-spend-too-much-time-on-phones/
86 https://www.dailymail.co.uk/news/article-4236684/Half-Europe-s-ready-meals-eaten-Britain.html
87 There have of course always been a small minority who love their work; increasing their number is an enduring utopian dream.
88 David Graeber, 'On the Phenomenon of Bullshit Jobs: A Work Rant', *Strike! Magazine*, Issue 3, August 2013: https://strikemag.org/bullshit-jobs
89 https://www.thetimes.co.uk/article/review-bullshit-jobs-a-theory-by-david-graeber-quit-now-your-job-is-pointless-9tk2l8jrq
90 See Chapter Two, p. 87.
91 https://www.about.sainsburys.co.uk/~/media/Files/S/Sainsburys/living-well-index/sainsburys-living-well-index-may-2018.pdf
92 Matthew Crawford, *The Case for Working with Your Hands: or Why Office Work is Bad for Us and Fixing Things Feels Good*, Penguin, 2009, p.2.
93 For a detailed discussion of the importance of the hand in human cognition, see Richard Sennett, *The Craftsman*, Penguin, 2009, pp.149–78.
94 http://www.economist.com/news/china/21631113-why-so-many-chinese-children-wear-glasses-losing-focus
95 It appears in Dryden's 1672 play *The Conquest of Granada*.
96 BedZED was designed by the architect Bill Dunster in collaboration with the eco-charity Bioregional: https://www.bioregional.com/projects-and-services/case-studies/bedzed-the-uks-first-large-scale-eco-village
97 https://journals.sagepub.com/doi/pdf/10.1177/0956247809339007

4 Society

1 Quoted in Ernest Mignon, *Les Mots du Général*, Librairie Arthème Fayard, 1962, Ch. 3.

2   https://www.rungisinternational.com/wp-content/uploads/2018/06/RUNGIS-RA_2017_EN_OK.pdf

3   For a detailed discussion of the relationship between markets and cities, see Steel, op. cit., pp. 105–52.

4   The glass and iron halls were built on the historic market site to designs by Victor Baltard in the 1850s.

5   Émile Zola, *The Belly of Paris* (*Le Ventre de Paris*, 1873), Brian Nelson (trans.), Oxford World Classics, OUP, 2007, p. 14.

6   See Stephen Kaplan, *Provisioning Paris: Merchants and Millers in the Grain and Flour Trade During the Eighteenth Century*, Ithaca and London, Cornell University Press, 1984.

7   The market porters, or *forts*, were instrumental in stirring unrest in the lead-up to the revolution.

8   Rungis opened in 1969.

9   Quoted in Kaplan, op. cit., p. 119.

10 Between 1858 and 1867 wheat prices in Chicago soared from 55 cents a bushel to $2.88 per bushel, before falling back to 77 cents. See Niall Ferguson, *The Ascent of Money: A Financial History of the World*, Penguin, 2009, p. 227.

11 http://triplecrisis.com/food-price-volatility/

12 Previously carried out by the US Commodity Futures Trading Commission.

13 Olivier de Schutter, *Food Commodities Speculation and Food Price Crises: Regulation to reduce the risks of price volatility*, UNFAO Briefing Note 02, September 2010, pp. 2–3.

14 Libor stands for London Interbank Offered Rate.

15 United Nations Conference on Trade and Development, *Key Statistics and Trends in International Trade 2014*, p. 7.; https://news.virginia.edu/content/global-food-trade-may-not-meet-all-future-demand-uva-study-indicates

16 De Schutter, op. cit., p. 1.

17 For a detailed discussion of how the development of public space has been shaped by food, see Steel, op. cit. pp. 118–33.

18 A point forcefully made by Carol Cadwalladr, the British journalist who uncovered the scandal: https://www.ted.com/talks/carole_cadwalladr_facebook_s_role_in_brexit_and_the_threat_to_democracy?language=en

19 Harari, op. cit., p. 27.

20 Rousseau, *The Social Contract and the First and Second Discourses*, p. 164.

21 Ibid., p. 166.

22 Shalom H. Schwartz, 'Value orientations: Measurement, antecedents and consequences across nations', in R. Jowell, C. Roberts, R. Fitzgerald, and G. Eva (eds.), *Measuring attitudes cross-nationally: lessons from the European Social Survey*, Sage, 2006.

23 Thomas Paine, *Rights of Man, Common Sense and Other Political Writings (1791)*, Oxford World Classics, OUP, 2008, p. 5.

24 Wrangham, op. cit., p. 133.

25 Dunbar found that the brain size of primates correlated directly to the number of individuals in any given group.

26   See Diamond, *The World Before Yesterday*, pp.12–20.

27   http://www.britannica.com/topic/slavery-sociology

28   Aristotle, *The Politics*, p.69.

29   The full quote is 'No one pretends that democracy is perfect or all-wise. Indeed it has been said that democracy is the worst form of Government except for all those other forms that have been tried from time to time.' Winston Churchill, House of Commons, 11 November 1947, https://api.parliament.uk/historic-hansard/commons/1947/nov/11/parliament-bill

30   Thomas Hobbes, *Leviathan* (1651), Cambridge University Press, 2004, p.87.

31   Ibid., p.33.

32   The term 'state of nature' appears to have been invented by Hobbes, although it was derived from the work of Grotius, who first spoke of humanity's 'natural laws and rights': ibid., p.xxviii.

33   Ibid., p.87.

34   Ibid., p.89.

35   Hobbes would probably have been unsurprised to learn that, ten years after the US-led invasion of Iraq, many Iraqis, weary of the resultant insurgency and chaos, longed for a return to the certainty of life under the despotic Saddam Hussein.

36   Hobbes, op. cit., p.120.

37   Although in fact Locke's *Treatises* were written explicitly to refute Robert Filmer's 1680 treatise *Patriarcha, or the Natural Power of Kings*. See John Locke, *Two Treatises of Government*, Peter Laslett (ed.), Cambridge University Press, 2015, pp.67–79.

38   Ibid., p.271.

39   Ibid., p.286.

40   Ibid., p.287.

41   Ibid., p.291.

42   Ibid., p.295.

43   Ibid., p.296.

44   Ibid., p.294.

45   Ibid., p.330.

46   Peter J. Hatch, *A Rich Spot of Earth: Thomas Jefferson's Revolutionary Garden at Monticello*, Yale University Press, 2012, p.3.

47   American Declaration of Independence, drafted by Thomas Jefferson with amendments by John Adams and Benjamin Franklin, ratified by Congress on 4 July 1776.

48   http://www.theguardian.com/business/2014/nov/13/us-wealth-inequality-top-01-worth-as-much-as-the-bottom-90

49   Locke, op. cit., p.293.

50   Marcel Mauss, *The Gift* (1950), Routledge, 2006, p.105.

51   See Branislow Malinowski, *Argonauts of the Western Pacific* (1922), Routledge, 2014.

52   Mauss, op. cit., pp.25–6.

53 See Evan D. G. Fraser and Andrew Rimas, *Empires of Food: Feast, Famine and the Rise and Fall of Civilizations*, Random House, 2010, pp.104–7.

54 Reay Tannahill, *Food in History*, Penguin, 1988, p.47.

55 See Ferguson, op. cit., p.31.

56 Ibid., pp.26–7.The Spanish committed this mistake in the sixteenth century, mining so much silver in the New World that its value collapsed at home.

57 Xenophon, *Ways and Means*, 4:7, quoted in Tomas Sedlacek, *Economics of Good and Evil: The Quest for Economic Meaning from Gilgamesh to Wall Street*, Oxford University Press, 2013, p.104.

58 Ibid., p.35.

59 'Capitalist' is first recorded in seventeenth-century Holland. See Fernand Braudel, *Civilization and Capitalism 15th–18th Century*, Vol. 2, Fontana, 1985, p.234.

60 Ibid., p.51.

61 Adam Smith, *The Wealth of Nations* Books I–III (1776), Penguin Classics, 1999, p.479.

62 Much of the activity was promoted by leading agronomists such as Arthur Young and Charles 'Turnip'Townshend. See Huggett, op. cit., p.66.

63 E. A.Wrigley, *Cities, People and Wealth: The Transformation of Traditional Society*, Blackwell, 1987, p.142.

64 Ibid.

65 Daniel Defoe, *Complete Tradesman*, ii, Ch. 6, quoted in George Dodd, *The Food of London*, Longman, Brown, Green and Longmans, 1856, pp.110–11.

66 For a detailed discussion of this problem, see Kaplan, op. cit.

67 Food shortages in Paris were a key element in the lead-up to the French Revolution. See Kaplan, op. cit.

68 Smith, op. cit., p.479.

69 Ibid., p.112. While a single workman 'could scarce make one pin a day' thanks to the numerous different tasks involved, said Smith, ten factory workers specialising in each task could between them churn out 48,000 pins in the same time.

70 Ibid., p.119.

71 Ibid., p.269.

72 Ibid., p.126.

73 See, for example, J. K. Galbraith, *The Affluent Society*; Amartya Sen, *Development as Freedom*; Joseph Stiglitz, *The Price of Inequality*; Thomas Piketty, *Capitalism in the Twenty-First Century*; also Tim Jackson, *Prosperity Without Growth* and Raj Patel, *The Value of Nothing*.

74 Adam Smith, *The Theory of Moral Sentiments* (1759), Penguin, 2009, p.13.

75 Ibid., p.73.

76 Ibid., p.213.

77 Ibid., pp.220, 221.

78 Ibid., p.19.

79 Karl Polanyi, *The Great Transformation: The Political and Economic Origins of Our Time* (1944), Beacon Press, 2001, p.45.

80 Ibid., p.44.

81 Ibid., p.171.

82 Karl Marx and Friedrich Engels, *The Communist Manifesto* (1848), Samuel Moore (trans.), Penguin, 1967, p.223.

83 Ibid., p.222.

84 Ibid., p.223.

85 Ibid., p.227.

86 Benjamin Franklin, 'Advice to a Young Tradesman, Written by an Old One', quoted by Max Weber in *The Protestant Ethic and the 'Spirit' of Capitalism* (1905), Peter Baehr and Gordon C. Wells (trans.), Penguin, 2002, p.9.

87 For a detailed discussion of this problem, see Simon Schama, *The Embarrassment of Riches: An Interpretation of Dutch Culture in the Golden Age*, Fontana Press, 1991.

88 The doctrine of predestination in Calvinism decrees that God decides whether one is destined for salvation or not before one is born.

89 For a discussion of the origins of capitalism, see Braudel, op. cit., pp.232–49.

90 Weber, op. cit., p.9.

91 Ibid. p.10.

92 Ibid., p.12.

93 Friedrich Hayek, *The Road to Serfdom* (1944), Routledge, 2001, p.13.

94 Ibid., p.14.

95 Ibid., p.13.

96 John Kenneth Galbraith, *The Affluent Society* (1958), Penguin, 1999, p.1.

97 The Bretton Woods Agreement to regulate monetary relations after the Second World War was made in 1944 between the forty-four Allied nations. The agreement led to the formation of the International Monetary Fund and the World Bank.

98 https://www.theguardian.com/business/2018/aug/16/ceo-versus-worker-wage-american-companies-pay-gap-study-2018

99 https://www.trusselltrust.org/news-and-blog/latest-stats/end-year-stats/

100 For a detailed discussion of global trade abuses, see Joseph Stiglitz, *Globalisation and its Discontents*, Penguin, 2002, pp.3–22.

101 Figures according to the National Audit Office: see https://www.nao.org.uk/highlights/taxpayer-support-for-uk-banks-faqs/

102 Joseph Stiglitz, *The Price of Inequality*, Penguin, 2013, p.40.

103 Michael Sandel, *What Money Can't Buy: The Moral Limits of Markets*, Penguin, 2013.

104 Galbraith, op. cit., p.66.

105 Alfred Marshall's *Principles of Economics*, which was the accepted textbook on economics for generations, in fact built on the work of others, notably the French economist Léon Walras and the English economist William Jevons.

106 https://www.ft.com/content/2ce78f36-ed2e-11e5-888e-2eadd5fbc4a4; https://www.cia.gov/library/publications/the-world-factbook/fields/2012.html

107 http://www.chinalaborwatch.org/reports

108   http://www.chinalaborwatch.org/upfile/2013_7_29/apple_s_unkept_promises.pdf; https://www.theguardian.com/global-development/2015/jul/20/thai-fishing-industry-implicated-enslavement-deaths-rohingya

109   https://www.theguardian.com/business/2016/jul/22/mike-ashley-running-sports-direct-like-victorian-workhouse

110   Carl Benedikt Frey and Michael A. Osborne, *The Future Of Employment: How Susceptible Are Jobs To Computerisation?*, Oxford Martin Programme on Technology and Employment, 17 September 2013: http://www.oxford martin.ox.ac.uk/publications/view/1314; 'When Robots Steal Our Jobs', *Analysis*, BBC Radio 4, 8 March 2015.

111   https://www.ons.gov.uk/employmentandlabourmarket/peopleinwork/employmenta ndemployeetypes/articles/whichoccupationsareathighestrisk ofbeingautomated/2019-03-25

112   E. F. Schumacher, *Small is Beautiful: A Study of Economics as if People Mattered* (1973), Vintage, 1993, p.2.

113   Rutger Bregman, *Utopia for Realists*, Bloomsbury, 2017.

114   Ibid. p.46.

115   Ibid.

116   John Maynard Keynes, 'Economic Possibilities for Our Grandchildren': http://www.econ.yale.edu/smith/econ116a/keynes1.pdf

117   Ibid.

118   https://www.theatlantic.com/magazine/archive/2013/06/are-we-truly-overworked/309321/

119   Schumacher, op. cit., p.8.

120   Ibid., p.84.

121   Ibid., p.85.

122   Ibid., p.40.

123   Quoted in Naomi Klein, *The Shock Doctrine*, Penguin, 2008, p.6.

124   The US Farm Bill, originally drawn up in 1933 as part of Roosevelt's New Deal, is worth one trillion dollars over ten years. Since payments are based on farm size, most of the money goes to big agribusiness. In 2014, 10,000 of the biggest farmers received between $100,000 and $1 million in subsidies, while the bottom 80 per cent got just $5,000 on average. EU farm subsidies, which represent 40 per cent of the EU's total budget, also favour big farmers. A 2016 report by Greenpeace found that more than one in five of the top hundred UK recipients were members of aristocratic families, with sixteen recipients (including the Queen and a racehorse-owning Saudi prince) on the *Sunday Times* Rich List. Between them, the top hundred received £87.9 million in agricultural subsidies, more than the bottom 55,119 recipients in the single payment scheme combined. See https://newrepublic.com/article/116470/farm-bill-2014-its-even-worse-old-farm-bill; https://www.theguardian.com/environment/2016/sep/29/the-queen-aristocrats-and-saudi-prince-among-recipients-of-eu-farm-subsidies

125   https://sustainablefoodtrust.org/key-issues/true-cost-accounting/

126   Thomas Aquinas, *Summa Theologica*, IIa–IIae Q.66.A.7 Corpus, quoted in Sedlacek, op. cit., p.150.

127   *International Business Times*: http://www.ibtimes.com/us-spends-less-food-any-other-country-world-maps-1546945, accessed 22 June 2014.

128   See Carlo Petrini, *Slow Food Nation: Why Our Food should be Good, Clean and Fair*, Rizzoli, 2007, pp.93–143.

129   http://www.greenpeace.org/international/en/campaigns/agriculture/problem/Corporations-Control-Our-Food/

130   Woody Tasch, *Inquiries into the Nature of Slow Money: Investing as if Food, Farms, and Fertility Mattered*, Chelsea Green, 2008.

131   https://slowmoney.org/our-team/founder/

132   Ibid.

133   Aditya Chakrabortty, 'In 2011 Preston hit rock bottom. Then it took back control', *Guardian*, 31 January 2018: https://www.theguardian.com/commentisfree/2018/jan/31/preston-hit-rock-bottom-took-back-control

134   Begun after the Spanish Civil War, the Mondragón Corporation is a federation of cooperatives that is now the tenth-largest company in Spain, employing 80,000 owner-workers.

135   Ibid.

136   https://www.theguardian.com/politics/2018/nov/01/preston-named-as-most-most-improved-city-in-uk

137   https://truthout.org/video/thomas-piketty-the-market-and-private-property-should-be-the-slaves-of-democracy/

138   Since capitalism naturally creates inequality, says Piketty, we need to build redistribution directly into the system. This could be done, he suggests, by imposing a 'confiscatory' tax of 80 per cent on incomes of $500,000 a year or more. In addition, there should be a progressive global tax on private wealth, which would require transparency for all bank transactions and the international sharing of data. See Thomas Piketty, *Capitalism in the Twenty-First Century*, Belknap Press of Harvard University Press, 2014, pp.512–20.

139   Leviticus 25: 2–5.

140   See Sedlacek, op. cit., p.76.

141   Mauss, op. cit., p.47.

142   The right to save and sow seeds is key to this, see: http://vandanashiva.com

143   Voltaire, *Candide* (1758), Philip Littell (trans.), Boni & Liveright, New York, 1918, p.167.

144   See Herman Daly, *Beyond Growth*, Beacon Press, 1996, pp.31–44.

145   See p.279.

146   The world's happiest nations, including those of Scandinavia and the current world champion, Denmark, have some of the lowest income ratios on earth. See Richard Wilkinson and Kate Pickett, *The Spirit Level: Why equality is better for everyone,* Penguin, 2010.

147   The inhabited garden is the most popular theme in the utopian cannon, kicking off with the Garden of Eden itself. Thomas More's 1516 *Utopia* proposed a network of self-sufficient city-states inhabited by vegetable-growing

enthusiasts; Ebenezer Howard's 1902 *Garden Cities of To-morrow* was essentially More's *Utopia* with railways. Meanwhile, William Morris's 1890 *News from Nowhere* imagined London transformed into a bucolic paradise in which rosy-cheeked peasants plucked apricots from trees in Trafalgar Square.

5 City and Country

1   Ben Flanner, interviewed by the author on 30 January 2017.
2   Ibid.
3   See Anastasia Cole Plakias, *The Farm on the Roof*, Avery, 2016.
4   'Feeding Our Cities in the 21st Century', Soil Association 60th Anniversary Conference press release, 12 September 2005.
5   www.verticalfarm.com
6   http://nymag.com/news/features/30020/
7   http://www.plantlab.nl
8   http://aerofarms.com/technology/
9   The calculation does not, however, include the non-productive areas of the vertical farm.
10   http://aerofarms.com/story/
11   Attributed to William Gibson.
12   For Despommier's own rundown of the advantages of vertical farming, see Dickson Despommier, *The Vertical Farm, Feeding the World in the 21st Century*, Picador, 2010, pp.145–75.
13   http://growing-underground.com
14   From an interview with the author in February 2018.
15   Despommier, op. cit., p.215.
16   George Dodd, *The Food of London*, Longman Brown, Green and Longmans, London, 1856, pp.222–3.
17   In their 2001 Pig City project, the Dutch architects MVRDV proposed a series of high-rise luxury towers to house the Netherlands' 15 million pigs, arguing that the porkers would be better off in luxury high-rise 'flats' with open-air balconies than in the dark, cramped conditions in which most are currently kept. See https://www.mvrdv.nl/projects/181-pig-city
18   http://www.fao.org/fileadmin/user_upload/newsroom/docs/en-so-law-facts_1.pdf
19   http://uk.businessinsider.com/inside-aerofarms-the-worlds-largest-vertical-farm-2016-3?r=US&IR=T
20   Prices from Ocado and Liffe, March 2019.
21   http://www.newyorker.com/magazine/2017/01/09/the-vertical-farm
22   For a detailed discussion of this question – and an argument that cities came first – see Jane Jacobs, *The Economy of Cities*, Vintage, 1969.
23   This idea was first postulated by Johann Von Thünen, whose 1826 *The Isolated State* was the first work to analyse how the productive hinterlands of cities would naturally develop.

24 Fraser and Rimas, op. cit., p.107.

25 *The Epic of Gilgamesh*, Andrew George (trans.), Penguin, 1999, p.5

26 Ibid., p.14.

27 Plato, *Laws*, V.738e, in *The Collected Dialogues*, Edith Hamilton and Huntingdon Cairns (eds), Princeton, 1987, p.1323.

28 Ibid., p.1324.

29 Aristotle, *The Politics*, pp.105–6.

30 For my own map of Rome's food supplies see Steel, op. cit., p.74.

31 The term 'food miles' was coined by City University, London Professor of Food Policy Tim Lang to describe the distance our food has travelled before we eat it.

32 Rome was dubbed the first consumption city by the German sociologist Werner Sombart in *Der Moderne Kapitalismus*, Leipzig and Berlin, 1916, pp.142–3, quoted in Neville Morley, *Metropolis and Hinterland*, Cambridge University Press, 1996, p.18.

33 Commentators including Pliny muttered that Rome's decline had begun with its taste for such luxury foods, bemoaning the city's dependence on others for its sustenance. See Morley, op. cit., p.88.

34 St Cyprian, bishop of Carthage, wrote in AD 250, 'The world has grown old and does not remain in its former vigour. It bears witness to its own decline.' Quoted in Herbert Girardet, *Cities People Planet*, Wiley Academy, 2004, p.46.

35 Rising salt levels in the third millennium BCE forced farmers to swap their preferred crop of wheat for barley, while poets lamented the 'whiteness of the fields'. See J. N. Postgate, *Early Mesopotamia: Society and Economy at the Dawn of History*, London and New York, Routledge, 1994, p.181.

36 Plato, *Critias*, 111c, Hamilton and Cairns, op. cit., p.1216.

37 *Cultus*' many meanings included tillage, worship and civilisation; it is also the origin of the word 'cult'.

38 In his *Gallic Wars*, Julius Caesar noted that, of all his foes, 'the Belgae are the bravest, because they are farthest from the civilisation and refinement of our Province, and merchants least frequently resort to them and import those things which tend to effeminate the mind'. Caius Julius Caesar, *De Bello Gallico & Other Commentaries*, W. A. Macdevitt (trans.), Everyman's Library, 1929, Book 1.

39 Tacitus, *Germania*, Ch. 16, M. Hutton (trans.), London, William Heinemann, 1970, p.155.

40 For a detailed analysis of the fresco, see Maria Luisa Meoni, *Utopia and Reality in Ambrogio Lorenzetti's* Good Government, Firenze, Edizioni IFI, 2006.

41 For a discussion of the emergence of the medieval Italian city-state, see Henri Lefebvre, *The Production of Space*, Donald Nicholson-Smith (trans.), Blackwell 1998, pp.78–9, 277–8.

42 An era signalled by the inaugural run of the British Stockton to Darlington Railway on 27 September 1825.

43 Christopher Watson, 'Trends In World Urbanisation', in *Proceedings of the First International Conference on Urban Pests*, K. B. Wildey and W. H. Robinson (eds), Centre for Urban and Regional Studies, University of Birmingham, UK, 1993.

44 William Cronon, *Nature's Metropolis: Chicago and the Great West*, New York, W. W. Norton and Co., 1991, pp.216–17.

45 Ibid., p.225.

46 The original, pre-railways Porkopolis had been Cincinnati.

47 Cronon, op. cit., p.244.

48 JBS acquired Swift and Co. for $1.5 billion in 2007.

49 https://www.bbc.co.uk/news/world-latin-america-46327634

50 http://www.mightyearth.org/forests/

51 Despite being a protected habitat set up with funds from the World Bank, much of the park was owned by cronies of then-President Michel Temer, himself mired in a corruption scandal involving JBS. See https://www.theguardian.com/world/2017/may/18/brazil-explosive-recordings-implicate-president-michel-temer-in-bribery

52 https://www.theguardian.com/world/2019/aug/02/brazil-space-institute-director-sacked-in-amazon-deforestation-row

53 Cronon, op. cit., p.198.

54 https://www.un.org/development/desa/en/news/population/2018-revision-of-world-urbanization-prospects.html

55 Ibid.

56 http://www.economist.com/news/china/21640396-how-fix-chinese-cities-great-sprawl-china

57 For the early uses of writing, see Bruce G. Trigger, *Understanding Early Civilisations*, Cambridge University Press, 2007, pp.588–90.

58 Friedrich Engels, op. cit., p.52.

59 *The Fastest Changing Place on Earth*, BBC2, first broadcast on 5 March 2012.

60 'Pitfalls Abound in China's Push From Farm to City', *New York Times*, 13 July 2013: http://www.nytimes.com/2013/07/14/world/asia/pitfalls-abound-in-chinas-push-from-farm-to-city.html?pagewanted=all

61 William C. Sullivan and Chun-Yen Chang (eds.), 'Landscapes and Human Health' (special issue), *International Journal of Environmental Research and Public Health*, May 2017 (ISSN 1660-4601): http://www.mdpi.com/journal/ijerph/special_issues/landscapes

62 https://www.agclassroom.org/gan/timeline/1900.htm

63 Joel Dyer, *Harvest of Rage: Why Oklahoma City Is Only the Beginning*, Westview Press, 1997, p.4.

64 Pollan, *The Omnivore's Dilemma*, p.52.

65 Ibid., p.53.

66 The 1994 North American Free Trade Agreement (NAFTA) flooded Mexico with cheap US corn, a hybrid version of the Mexican staple that had until then been grown in forty varieties. The result was to push almost half of Mexico's farmers – 1.3 million of them – into cities, where they

found themselves paying more for food than they had done when they grew it themselves. See Raj Patel, *Stuffed and Starved: Markets, Power and the Hidden Battle for the World Food System*, Portobello, 2007, pp.48–54.

67 Ibid., p.3.

68 James T. Horner and Leverne A. Barrett, *Personality Types of Farm Couples*, University of Nebraska, 1987, quoted in ibid., p.35.

69 Ibid., p.19.

70 Doug Saunders, *Arrival City: How the Largest Migration in History is Shaping our World*, Windmill Books, 2011, p.1.

71 Ibid., p.121.

72 Ibid., p.122.

73 Ibid., p.128.

74 Ibid., p.112.

75 http://english.gov.cn/state_council/2014/09/09/content_281474986284089.htm

76 Stewart Brand, *Whole Earth Discipline*, Atlantic Books, 2009, p.44.

77 Ibid., p.47.

78 Ibid., p.39.

79 Diana Lee-Smith, 'My House is My Husband: A Kenyan Study of Women's Access to Land and Housing', PhD thesis for Lund University Sweden, 1997, pp.143–4.

80 Brand prefers the term 'genetic engineering' to 'genetic modification' because, as he points, out, all of evolution has involved the latter. See ibid., p.118.

81 Ibid., p.27.

82 Patel, *Stuffed and Starved*, pp.119–27.

83 Ibid., pp.126–7.

84 Ebenezer Howard, *Garden Cities of To-Morrow* (1902), MIT Press, 1965, p.48.

85 Ibid., pp.45–6.

86 Henry George, 'What the Railroad Will Bring Us', *The Overland Monthly*, Vol. 1, October 1868, No. 4, pp.297–306, https://quod.lib.umich.edu/m/moajrnl/ahj1472.1-01.004/293:1?rgn=full+text;view=image

87 Henry George, *Progress and Poverty* (1879), Pantianos Classics, 1905, p.107.

88 Residents would pay a special 'rate-rent' charge, in which the 'rent' would pay off the original loan, while the 'rate' would go towards public works and services such as healthcare and pensions, thus effectively creating a local welfare state. See Robert Beevers, *The Garden City Utopia: A Critical Biography of Ebenezer Howard*, Macmillan Press, 1988, p.62.

89 Ibid., p.14

90 Ibid., p.79.

91 Ibid., p.76.

92 For a detailed description of the building of Letchworth, see Peter Hall, *Cities of Tomorrow*, Blackwell, 2002, pp.97–101.

93 Greek *eu,* good + *topos,* place; or *ou,* no + *topos,* place.

Notes

94   Anna Minton, *Big Capital*, Penguin, 2017, p.3.

95   Ibid., p.54.

96   Ibid., p.xiv.

97   Rem Koolhaas, 'Countryside', *O32C*, Issue 23, Winter 2012/2013, pp.49–72.

98   Ibid., p.62.

99   Ibid., p.61.

100   Ibid., p.53.

101   Pliny the Younger, Ep. 2.17.2, quoted in Morley, op. cit., p.91.

102   Ironically, Howard intended his garden cities to be densely built, although the results were anything but.

103   Makoto Yokohari, 'Agricultural Urbanism: Re-designing Tokyo's Urban Fabric with Agriculture', Herrenhausen Conference, Hanover, May 2019: https://www.researchgate.net/publication/329999704_Agricultural_Urbanism_Re-designing_Tokyo's_Urban_Fabric_with_Agriculture_Preprint

104   Patrick Geddes, *Cities in Evolution* (1915), Routledge, 1997, p.96.

105   Kate Raworth, *Doughnut Economics: Seven Ways to Think Like a 21st Century Economist*, Random House, 2017.

106   Patrick Geddes, 'The Valley Plan of Civilization', *The Survey*, 54, pp.40–4, quoted in Peter Hall, op. cit., p.149.

107   Ibid., p.97.

108   A saying frequently attributed to Mark Twain.

109   Tim Lang, Erik Millstone and Terry Marsden, 'A Food Brexit: Time to Get Real', July 2017: https://www.sussex.ac.uk/webteam/gateway/file.php?name=foodbrexitreport-langmillstonemarsden-july2017pdf.pdf&site=25

110   Ibid., p.18.

111   Lizzie Collingham, *The Taste of War: World War Two and the Battle for Food*, Penguin, 2012, pp.90–1.

112   Rousseau, *The Social Contract and the First and Second Discourses*, p.113.

113   Pierre-Joseph Proudhon: 'What is Property?', in *Property is Theft! – A Pierre-Joseph Proudhon Anthology*, Iain McKay (ed.), A. K. Press, 2011, p.87.

114   Ibid., p.95.

115   Ibid., p.131.

116   Ibid., pp.130–1.

117   Ibid., p.136.

118   Ibid., p.136.

119   Ibid., p.137.

120   Peter Kropotkin, *The Conquest of Bread* (1892), Penguin, 2015, p.19.

121   Ibid., p.13.

122   George Orwell, *Homage to Catalonia* (1938), Penguin, 2000, pp.3–4.

123   Ibid., p.98.

124   Another more recent example is the Kurdish state of Rojava in north-east Syria, which has been working on anarchist principles since 2012, also under conditions of civil war. See the documentary *Accidental Anarchist: Life Without Government*, BBC4, 23 July 2017.

125 Quoted in Martin Buber, *Paths in Utopia* (1949), R. F. C. Hull (trans.), Boston, Beacon Press, 1958, p.42.

126 Peter Kropotkin, *Fields, Factories and Workshops, or, Industry Combined with Agriculture and Brain Work with Manual Work* (1898), Martino Publishing, 2014, p.5.

127 Ibid., p.7.

128 Ibid., p.38.

129 Ibid., p.21.

130 Ibid., p.180.

131 Ibid., p.217.

132 Henry George, *Progress and Poverty*, p.120.

133 As Noam Chomsky has argued, anarchism is indeed enjoying a revival today in the form of the Occupy movement. See Noam Chomsky, *On Anarchism*, Penguin, 2013.

134 http://www.bbc.co.uk/news/magazine-33133712

135 http://www.countrylife.co.uk/articles/who-really-owns-britain-20219

136 George, *Progress and Poverty*, p.147.

137 Ibid.

138 For an in-depth discussion of how land value taxes might work, see Martin Adams, *Land: A New Paradigm for a Thriving World*, North Atlantic Books, 2015. As Martin points out, one is not so much taxing land as asking landlords to recompense communities for the common resources from which they are being excluded; he therefore prefers the term 'community land contribution'.

139 George wasn't the first to posit the idea, however; that distinction goes to none other than Adam Smith, who first proposed a tax on ground rents on the basis that every landowner acts 'always as a monopolist, and exacts the greatest rent which can be got for the use of his ground'. Adam Smith, *The Wealth of Nations* (1776), Books IV–V, Penguin Classics, 1999, p.436.

140 https://www.theguardian.com/commentisfree/2017/oct/11/labour-global-economy-planet

141 Aristotle, *The Politics*, p.108.

142 The term was first coined by the British economist William Forster Lloyd in *Two Lectures on the Checks to Population*, Oxford University, 1833. See also Garrett Hardin, 'The Tragedy of the Commons', *Science*, Vol. 162, Issue 3859, 1968, pp.1243–8: http://science.sciencemag.org/content/162/3859/1243.full

143 Hardin used the example of a common grazing ground, which a rational herder would be compelled to overgraze on the basis that he would get direct benefit from the feeding of his own animals, but would only bear the shared cost of the overuse among the group. Hardin, op. cit., p.1244.

144 Garrett Hardin, 'Political Requirements for Preserving Our Common Heritage', *Wildlife and America*, H. P. Bokaw (ed.), Washington DC, 1978, p.314.

145 Elinor Ostrom, 'Beyond Markets and States: Polycentric Governance of Complex Economic Systems', *American Economic Review* 100, June 2010, p.10: http://www.aeaweb.org/articles.php?doi=10.1257/aer.100.3.1

146 Small to medium-sized cities, Ostrom found, were far better at monitoring their resources than larger ones.

147 Elinor Ostrom, 'Beyond Markets and States', talk given at Indiana University, 2009.

148 Raymond J. Struyk and Karen Angelici, 'The Russian Dacha Phenomenon', *Housing Studies*, Volume 11, Issue 2, April 1996, pp.233–50.

149 https://www.foodcoop.com

150 www.growingpower.org.

151 https://stephenritz.com/the-power-of-a-plant/

152 https://www.growingcommunities.org

153 See André Viljoen (ed.), *CPULs, Continuous Productive Urban Landscapes*, Architectural Press, 2005.

154 When back in the 2000s Rob Hopkins started the Transition Movement – in which towns and local groups collaborate to reduce their carbon emissions over time – food was just one of the items on his agenda; he quickly realised, however, that food-based projects were by far the most effective in getting people engaged and involved. (Conversation with the author, 2009.)

6 Nature

1 Wendell Berry, *Home Economics*, Counterpoint, Los Angeles, 1987, p.10.

2 'Anthropocene' was coined by Eugene F. Stoermer in the 1980s and popularised by Paul J. Crutzen.

3 Harari, op. cit., p.65.

4 Ibid., p.67.

5 Gerardo Ceballos, Paul R. Ehrlich, and Rodolfo Dirzo, 'Biological annihilation via the ongoing sixth mass extinction signaled by vertebrate population losses and declines', *PNAS*, 25 July 2017: http://www.pnas.org/content/114/30/E6089

6 Ibid.

7 http://journals.plos.org/plosone/article?id=10.1371/journal.pone.0185809

8 https://news.nationalgeographic.com/2018/05/farmland-birds-declines-agriculture-environnment-science/; https://www.independent.co.uk/environment/uk-bird-numbers-species-declines-british-wildlife-turtle-dove-corn-bunting-willow-tits-farmland-a7744666.html

9 https://www.birdlife.org/sites/default/files/attachments/BL_ReportENG_V11_spreads.pdf, pp.21, 23.

10 Ibid., p.31.

11 Warming oceans are not only destroying coral reefs – crucial habitats for marine biodiversity – but the resulting acidification is making it hard for coastal shellfish such as oysters to farm their shells properly.

12   https://www.theguardian.com/environment/2017/dec/14/a-different-dimension-of-loss-great-insect-die-off-sixth-extinction

13   https://www.bbc.com/news/uk-43051153

14   David R. Montgomery and Anne Biklé, *The Hidden Half of Nature: The Microbial Roots of Life and Health*, New York, W. W. Norton and Company, 2016, p.24.

15   Spector, op.cit., p.25.

16   Montgomery and Biklé, op. cit., p.2.

17   A 3.5-billion-year-old fossil found in Western Australia in 2013 is the oldest living thing yet discovered, a centimetre-thick colony of single-cell microbes formed around such a vent. See Seth Borenstein, 'Oldest Fossil Found: Meet Your Mom', 13 November 2013, Associated Press: http://apnews.excite.com/article/20131113/DAA1VSC01.html

18   https://www.scientificamerican.com/article/origin-of-oxygen-in-atmosphere/

19   Aristotle, *Physics*, Robin Waterfield (trans.), Oxford University Press, 1996, p.56.

20   Aristotle, *The Politics*, p.79.

21   See Chapter One, p. 28.

22   For a detailed discussion of this topic, see Ernst Cassirer, *The Individual and the Cosmos in Renaissance Philosophy*, University of Pennsylvania Press, 1963.

23   René Descartes, *Discourse on Method and Meditations on First Philosophy*, Donald A. Cress (trans.), Hackett, 1998, pp.18–19.

24   Ibid., p.31.

25   For a discussion of the problems associated with this abstract view, see Dalibor Vesely, *Architecture in the Age of Divided Representation*, MIT Press, 2004, pp.188–96.

26   Keith Thomas, *Man and the Natural World: Changing Attitudes in England 1500–1800*, Allen Lane, 1983, p.18.

27   Ibid., p.34.

28   See, for example, the writings of Temple Grandin, or Rosamund Young, *The Secret Life of Cows*, Faber & Faber, 2017.

29   https://www.vegansociety.com/about-us/further-information/key-facts

30   http://www.bbc.co.uk/news/uk-england-43140836

31   https://www.thebureauinvestigates.com/stories/2018-01-30/a-game-of-chicken-how-indian-poultry-farming-is-creating-global-superbugs

32   See Jules Pretty, *Agri-culture – Reconnecting People, Land and Nature*, Earthscan, 2002, pp.126–45.

33   In 2018, glyphosate hit the headlines for an even more sinister reason, when a groundskeeper, Dewayne Johnson, successfully sued Monsanto for damages after a court found that working regularly with the company's weed-killer RangerPro had contributed 'substantially' to his having contracted non-Hodgkin's lymphoma, a form of terminal cancer.

34   'Phytochemical' is from the Greek *phyton* – plant.

35 Figures given by Professor Ralph C. Martin in a lecture at the Toronto Food Policy Council, October 2011.: https://www.plant.uoguelph.ca/rcmartin

36 See Philippe Descola, *Beyond Nature and Culture*, Janet Lloyd (trans.), University of Chicago Press, 2013.

37 Ibid., p.46.

38 Jean-Jacques Rousseau, *The Social Contract and Discourses*, G. D. H. Cole (trans.), London and Toronto, J. M. Dent and Sons, 1923, p.145.

39 Rousseau, *The Social Contract and the First and Second Discourses*, p.48.

40 Ibid., p.52.

41 Henry David Thoreau, 'Walking', *The Works of Thoreau*, Henry S. Canby (ed.), Boston, Houghton Mifflin, 1937, p.672.

42 Ralph Waldo Emerson, *Nature* (1836), in *Nature and Selected Essays*, Penguin, 2003, p.35.

43 Ibid., p.38.

44 Ibid., p.37.

45 Ibid., p.39.

46 Ibid., p.43.

47 Henry David Thoreau, *Walden, or Life in the Woods* (1854), Oxford University Press, 1997, p.122.

48 John Muir, *A Thousand-mile Walk to the Gulf*, Boston and New York, Houghton Mifflin, 1916, p.xxxii.

49 John Muir, The Yosemite: http://www.gutenberg.org/files/7091/7091-h/7091-h.htm

50 John Muir, 'The Treasures of the Yosemite' and 'Features of the Proposed Yosemite National Park', *The Century Magazine*, Vol. 40, No. 4, August 1890, and No. 5, September 1890.

51 Ralph Waldo Emerson, op. cit., p.80.

52 John Muir, *My First Summer in the Sierra*, Boston and New York, Houghton Mifflin, 1911, p.99.

53 William Cronon, 'The Trouble with Wilderness', in William Cronon (ed.), *Uncommon Ground: Rethinking the Human Place in Nature*, New York, W. W. Norton and Company, 1996, p.80.

54 Wendell Berry, *Home Economics*, Counterpoint, Los Angeles, 1987, p.11.

55 Ibid., p.139.

56 Ibid., pp.7–8.

57 Ibid., p.6.

58 Ibid., p.140.

59 Ibid., p.142.

60 Ibid., p.143.

61 https://www.express.co.uk/life-style/food/694191/cheese-UK-Britain-France-brie-cheddar-producer-world-Monty-Python-Cathedral-City; https://www.cheesesociety.org/industry-data/

62 *Back to the Land*, Series 2 Episode 9, BBC2, 21 May 2018.

63 UNFAO, 'The Future of Food and Agriculture, Trends and Challenges', p.x.

64 Ibid., p.xi.
65 Ibid., p.7.
66 Ibid., p.xi.
67 The term 'permaculture' (from permanent agriculture) was invented by Bill Mollison and David Holmgren. See *Permaculture One*, Transworld, 1978.
68 UNFAO, op.cit., pp.48–9.
69 Ibid., p.5.
70 Ibid., p.1.
71 Ibid., p.9.
72 Fairlie includes just 56 grams of meat and 568 grams of dairy, basing his figures on a 1975 study by Scottish ecologist Kenneth Mellanby. See Simon Fairlie, *Meat: A Benign Extravagance*, Chelsea Green, 2010, p.95.
73 Ibid, pp.95–7.
74 The outbreak was traced to the feeding of some improperly rendered contaminated pigswill to pigs, a recurrence of which could be avoided with proper regulation. A UK group called the Pig Idea is fighting to overturn the ban, which was also imposed in the EU the following year, arguing that the waste of such a valuable resource – widely used in the rest of the world – makes no sense. See http://www.thepigidea.org
75 Quoted in Fairlie, op. cit., p.29.
76 Ibid., p.21.
77 Ibid., p.2.
78 Ibid., p.42.
79 Ibid., pp.38–9.
80 Ibid., p.40. Since farming this way would produce a grain surplus of around 150 million tonnes, Fairlie reckons we could afford to feed some grain to omnivorous beasts in addition, thus providing ourselves with an additional eight kilos of meat per person and a 'food buffer' against a lean harvest.
81 https://www.savory.global/our-mission/; https://atlasofthefuture.org/futurehero-tony-lovell-5-billion-hectares-hope/
82 Fairlie, op. cit., p.172. The figure is based on the reoccupation rates found in parts of the Serengeti.
83 Ibid, p.171. Rice accounted for an estimated 10 per cent of global methane emissions in 1990 – more than meat and dairy combined – a figure now thought to have been reduced by two thirds since China replaced organic manures with chemical fertilisers, although where emissions from the 'spare' organic manure – quantities of which have greatly expanded – is now being counted is unclear.
84 Adrian Muller, Christian Schader, Nadia El-Hage Scialabba, Judith Brüggemann, Anne Isensee, Karl-Heinz Erb, Pete Smith, Peter Klocke, Florian Leiber, Matthias Stolze and Urs Niggli, 'Strategies for Feeding the World more Sustainably with Organic Agriculture', *Nature Communications*, Vol.8, Article No. 1290, 2017: http://www.nature.com/articles/s41467-017-01410-w

85   Ibid., p.3. See also Simon Fairlie, where he points out that the huge discrep-
      ancies in UK wheat yields compared to other grains are almost certainly due
      to the lack of investment in organic varieties. Fairlie, op. cit., pp.87–8,

86   Tom Bawden, 'Organic farming can feed the world if done right, scientists
      claim', *Independent*, 10 December 2014: https://www.independent.co.uk/
      environment/organic-farming-can-feed-the-world-if-done-right-scien-
      tists-claim-9913651.html

87   J. N. Pretty, J. I. L. Morison and R. E. Hine, 'Reducing food poverty by
      increasing agricultural sustainability in developing countries', *Agriculture,
      Ecosystems and Environment* 95, 2003, pp.217–34.

88   http://www.worldwatch.org/node/4060

89   https://eatforum.org/eat-lancet-commission/

90   https://www.ipcc.ch/report/srccl/

91   Albert Howard, *An Agricultural Testament* (1940), Oxford University Press,
      1956, p.1.

92   Another reason why vertical farms in cities are not the answer is that they
      import fertility from far afield and fail to return it to the soil.

93   Howard, op. cit., p.4.

94   The obsession was fuelled by the spectacular gains brought about by the
      dumping of guano on the land after the Peruvian bird poo's miraculous
      properties (well known to the Incas) were discovered by the Prussian
      geographer-explorer Alexander von Humboldt in 1804, marking
      the start of its use in North America and Europe. See Smil, op. cit.,
      pp.39–42.

95   Howard, op. cit., p.161.

96   Ibid., pp.56–62, 166–8.

97   Ibid., p.18.

98   Ibid., pp.42–3.

99   Ibid., p.22.

100   Ibid., p.27.

101   Ibid.

102   Eve Balfour was the author of another organic farming classic, the 1943
      book *The Living Soil*. William Albrecht, president of the Soil Science Society
      of America, was also a Howard fan.

103   Montgomery and Biklé, op. cit., p.104.

104   Ibid., p.105.

105   Ibid., p.100.

106   Ibid., p.99.

107   Masanobu Fukuoka, 'The One-Straw Revolution', *New York Review of
      Books*, 1978, p.45.

108   Ibid., p.3.

109   In Liebig's last book, *The Natural Laws of Husbandry*, written in 1863, he
      overturned his earlier assumptions and stated that organic matter should be
      returned to the fields. See David R. Montgomery, *Growing a Revolution:
      Bringing Our Soil Back to Life*, W. W. Norton, 2017, pp.246–9.

110 Eve Balfour, *The Living Soil* (1943), Soil Association, 2006, p.21.

111 https://microbiomejournal.biomedcentral.com/articles/10.1186/s40168-017-0254-x

112 The most common complex carbohydrate is cellulose, present in almost all plants and responsible for their structure and flexible strength.

113 Kellogg's habit of shoving yoghurt up his patients' rear ends turns out not to have been quite as batty as it sounds.

114 Spector, op. cit., p.18.

115 The trip was recorded by Dan Saladino of BBC Radio 4's *Food Programme* in two episodes entitled 'Hunting with the Hadza', broadcast on 3 and 10 July 2017: https://www.bbc.co.uk/programmes/b08wmmwq

116 Ibid.

117 Michel de Montaigne, *The Complete Works*, Donald M. Frame (trans.), Everyman's Library, p.385.

118 See Alex Laird, *Root to Stem, A Seasonal Guide to Natural Recipes and Remedies for Everyday Life*, Penguin 2019, pp.63–6.

119 Dan Barber, *The Third Plate: Field Notes on the Future of Food*, Abacus, 2014, p.7.

120 Ibid., p.8.

121 Ibid.

122 Isabella Tree, *Wilding: The Return of Nature to a British Farm*, Picador, 2018, p.8.

123 For a discussion of the benefits of rewilding, see George Monbiot, *Feral: Rewilding the Land, Sea and Human Life,* Penguin 2014.

124 Qing Li, *Effect of forest bathing trips on human immune function,* Japanese Society for Hygiene, 2009: https://link.springer.com/article/10.1007/s12199-008-0068-3

125 For a summary of the latest such research, see, for example, http://www.julespretty.com/research/nature-and-health/

7 Time

1 *Dying to Talk*, BBC World Service, 27 April 2017: https://www.bbc.co.uk/programmes/p0506ttc

2 https://deathcafe.com

3 Death Café is now run by Underwood's mother and sister, Susan Barsky Reid and Jools Barsky.

4 Dylan Thomas, 'Do not go gentle into that good night', *The Poems of Dylan Thomas*, New Directions, 1971, p.239.

5 https://faithsurvey.co.uk/download/uk-religion-survey.pdf

6 https://www.nbcnews.com/better/wellness/fewer-americans-believe-god-yet-they-still-believe-afterlife-n542966

7 Atul Gawande, *Being Mortal: Illness, Medicine and What Matters in the End*, Profile Books, 2014, p.1.

8   Ibid., p.155.

9   Ibid., p.178.

10   Ibid., p.177.

11   Ibid., p.178.

12   https://www.theguardian.com/society/2017/sep/27/rise-in-uk-life-expectancy-slows-significantly-figures-show

13   *Eat to Live Forever with Giles Coren*, BBC2, 18 March 2015.

14   https://nutrisci.wisc.edu/richard-weindruch/

15   https://www.thetimes.co.uk/article/tv-review-eat-to-live-forever-with-giles-coren-the-billion-dollar-chicken-shop-6nz30cn7jgr

16   See Bernard Williams, 'The Makropulos case: reflections on the tedium of immortality', *Problems of the Self*, Cambridge University Press, 1973. https://web.archive.org/web/20160528020748/http://stoa.org.uk/topics/death/the-makropulos-case-reflections-on-the-tedium-of-immortality-bernard-williams.pdf

17   *The Epic of Gilgamesh*, op. cit., p.86.

18   Sarah Harper, *How Population Change Will Transform Our World,* Oxford University Press, 2016, p.2.

19   According to the NHS, the cost of looking after an eighty-year-old is five times that of looking after a thirty-year-old. See https://www.england.nhs.uk/five-year-forward-view/next-steps-on-the-nhs-five-year-forward-view/the-nhs-in-2017/#two

20   Ibid.

21   https://www.independent.co.uk/news/world/asia/india-railway-jobs-apply-recruitment-porters-cleaners-track-maintainers-unemployment-a8714016.html

22   Andrew Marvell, 'To his Coy Mistress', *The Metaphysical Poets*, Penguin, 1972, p.252.

23   Martin Nillson, *Primitive Time Reckoning*, Oxford University Press, 1920, p.42, quoted in Ingold, op. cit., p.325.

24   E. E. Evans-Pritchard, *The Nuer*, Oxford University Press, 1940, p.103.

25   Lewis Mumford, *Technics and Human Development, The Myth of the Machine*, Vol. 1, Harvest-HBJ, 1966, p.286.

26   See the earlier discussion of E. F. Schumacher's thinking in Ch. Four, p.160.

27   For those who struggle to get up in the morning, it may be comforting to know that there is a range of such personal rhythms, from 23.8 to 24.8 hours, which explains why some of us are 'sparrows', programmed to leap out of bed at dawn, and some – like me – are late-night 'owls'.

28   https://www.independent.co.uk/life-style/health-and-families/health-news/adults-uk-under-sleeping-health-sleep-fatigue-a6963631.html

29   https://www.ncbi.nlm.nih.gov/pmc/articles/PMC3763921/

30   Charles H. Kahn, *The Art and Thought of Heraclitus*, Cambridge University Press, 1979, p.53.

31   Hesiod, op. cit., p.40.

32 Ibid., p.42.

33 The second law of thermodynamics emerged from the work of the nineteenth-century physicists Sadi Carnot and Rudolf Clausius.

34 Seneca, *Dialogues and Essays*, John Davie (trans.), Oxford University Press, 2008, p.75.

35 https://www.worldlifeexpectancy.com/country-health-profile/india

36 The figure in the USA was 46 per cent. See https://www.gapminder.org/data/

37 https://www.historytoday.com/jared-bernard/dreaded-sweat-other-medieval-epidemic

38 The Stoa Poikile (Painted Stoa), was a long colonnaded public lounge fringing the agora in which everyone was free to gather.

39 The Stoics were founded around 300 BCE by Zeno of Citium, but no writings of his survive.

40 Philodemus, Herculaneum Papyrus, 1005, 4.9.14.

41 Seneca, op. cit., p.10.

42 Ibid., Letter LXI.4, p.427.

43 Ibid., Letter XII.4. p.67.

44 Ibid., Vol. 3, Letter XCIX.10, p.135.

45 According to Tacitus, Seneca died so slowly that he had to take hemlock – the same poison that Socrates had used – and immerse himself in a warm bath in order to finish the job. *The Annals of Tacitus*, John Jackson (trans.), Loeb Classical Library, 1937, Vol. V, Book XV.

46 With respect to this, it is revealing to ask yourself the following question: would you rather live without the Internet or without a flushing toilet?

47 Gawande, op. cit., p.19.

48 http://www.un-documents.net/our-common-future.pdf

49 Proudhon, op. cit., p.93.

50 I am grateful to Per Kølster for this reference, which is from the Danish organic farmer Eskild Rommer.

51 See Mikhail Bakhtin, *Rabelais and His World*, Hélène Iswolsky (trans.), MIT Press, 1968, p.6.

52 'Carnival': from Latin *carnis*, meat + *levare*, to put away.

53 Bakhtin, op. cit., p.6.

54 Ibid., p.10.

55 Ibid., p.21.

56 Ibid., p.27.

57 Timothy Morton, *Dark Ecology: For a Logic of Future Coexistence*, Columbia University Press, 2018, p.24.

58 Ibid., p.118, p.123.

59 Ibid., p.75.

60 Ibid., p.69.

61 http://www.joannamacy.net

62 Carlo Rovelli, *The Order of Time*, Allen Lane, 2018, pp.86–7.

63   Ibid., p.92.

64   See Ch. One, p.34.

65   Ingold, op. cit., p.331.

66   Aldous Huxley, *Brave New World* (1932), Vintage, 2007, p.47.

67   https://edition.cnn.com/2017/09/18/health/opioid-crisis-fast-facts/index.
      html

68   Rainer Maria Rilke, 'Sonnet XXVII', *In Praise of Mortality: Selections from
      Rainer Maria Rilke's Duino Elegies and Sonnets to Orpheus*, Anita Barrows and
      Joanna Macy (trans.), Riverhead Books, 2005, p.133.

69   Emerson, op. cit., p.44.

70   The concept originally came from work done by gerontologists Gianni Pes
      and Michel Poulain, who identified Sardinia's Nuoro province as a region
      with the highest concentration of male centenarians in the world, and drew
      concentric blue circles on a map centred on the area with the highest dens-
      ity: the Blue Zone.

71   https://www.bluezones.com

72   Okinawans, meanwhile, follow the 2500-year-old Confucian mantra '*Hara
      hachi bu*,' the '80-per-cent-rule', which reminds diners to stop eating when
      their stomachs are 80 per cent full.

73   Daniel Klein, *Travels with Epicurus*, Penguin, 2012, pp.22–3.

74   Marcus Aurelius, *Meditations*, Book 6.37, Penguin, 2006, p.53.

75   Mircea Eliade, *The Sacred and the Profane: The Nature of Religion*, Willard R.
      Trask (trans.), Harvest/Harcourt Brace Jovanovich, 1959, p.68.

Index

Index

penguin.co.uk/vintage